杨振宁
的三篇学位论文

Chen Ning Yang's Theses With Commentary

朱邦芬 阮东 编

清华大学出版社
北京

版权所有，侵权必究。举报: 010-62782989, **beiqinquan@tup.tsinghua.edu.cn**。

图书在版编目(CIP)数据

杨振宁的三篇学位论文/朱邦芬, 阮东编.—北京：清华大学出版社，2022.9
ISBN 978-7-302-61649-8

Ⅰ.①杨⋯　Ⅱ.①朱⋯　②阮⋯　Ⅲ.①物理学－文集　Ⅳ.①O4-53

中国版本图书馆 CIP 数据核字(2022)第 145518 号

责任编辑：魏贺佳
封面设计：何凤霞
责任校对：王淑云
责任印制：曹婉颖

出版发行：清华大学出版社
　　　网　　址：http://www.tup.com.cn, http://www.wqbook.com
　　　地　　址：北京清华大学学研大厦 A 座　　　邮　　编：100084
　　　社 总 机：010-83470000　　　邮　　购：010-62786544
　　　投稿与读者服务：010-62776969, c-service@tup.tsinghua.edu.cn
　　　质量反馈：010-62772015, zhiliang@tup.tsinghua.edu.cn
印 装 者：北京博海升彩色印刷有限公司
经　　销：全国新华书店
开　　本：185mm×260mm　　　印　张：11.5　　　字　数：254 千字
版　　次：2022 年 9 月第 1 版　　　印　次：2022 年 9 月第 1 次印刷
定　　价：98.00 元

产品编号：078177-01

谨以本书庆贺
杨振宁先生百岁寿辰

西南联大硕士研究生毕业时的杨振宁 (1944)

芝加哥大学博士研究生毕业时的杨振宁 (1948)

目　　录

Comments on My Three Theses .... Chen Ning Yang / 1

Group Theory and the Vibration of Polyatomic Molecules
（学士学位论文，1942 年）.. 5

Investigations in the Statistical Theory of Superlattices
（硕士学位论文，1944 年）.. 43

On the Angular Distribution in Nuclear Reactions and Coincidence Measurements
（博士学位论文，1948 年）.. 103

杨振宁在国立西南联合大学期间的学业 146

杨振宁的国立西南联合大学及芝加哥大学成绩单 149

西南联大的人才培养和杨振宁先生的学术起步
——代后记 .... 朱邦芬 / 153

Comments on My Three Theses

Chen Ning Yang

Professors B.F. Zhu and R. Dong proposed to publish the 3 theses which I wrote in the 1940s. Each of the three was very important for my later research work, important in very different ways. So I thought detailing their respective influence on me may be useful for graduate students at the beginning of their careers.

1942 thesis for my BSc degree

At the Southwestern Associated University where I matriculated, 1938 to 1942, students were required to submit a thesis for their BSc degree. I had taken a course in quantum theory from Professor T. Y. Wu (吴大猷). So I went to him asking him to be my supervisor. What then happened I had described, as follows[1]:

With Professor Wu, 1982 in StonyBrook.

> he gave me a copy of an article by J. E. Rosenthal and G. M. Murphy in the 1936 volume of Reviews of Modern Physics. It was a review paper on group theory and molecular spectra. I was thus introduced to group theory in physics. In retrospect I am deeply grateful to Wu for this introduction, since it had a profound effect on my subsequent development as a physicist.

The importance to me of this early introduction to group theory's role in physics cannot be overemphasized. And I was deeply aware of this already in the early 1950s. In October 1957, one day after I learned I was to receive the Nobel Prize jointly with T.D. Lee, I wrote to Professor Wu[2]:

---

[1] C.N. Yang. Selected Papers 1945–1980 With Commentary. Freeman and Company (1983), p. 5.

[2] C.N. Yang. Selected Papers 1945–1980 With Commentary. Freeman and Company (1983), p. 41.

At this moment of great excitement, that also calls for deep personal reflection, it is my privilege to express to you my deep gratitude for your having initiated me into the field of symmetry laws and group theory in the spring of 1942. A major part of my subsequent work, including the parity problem, is traceable directly or indirectly to the ideas that I learned with you that spring fifteen years ago. This is something that I have always had an urge to tell you, but today is a particularly appropriate moment.

1944 thesis for my MSc degree

This thesis was written under the direction of Professor J. S. Wang (王竹溪). It consisted of two papers on the specific heat of alloys, using approximations which were very popular at the time. The main ideas of such approximations are now called mean field theory. Neither paper made any impact in the field, but they did introduce me to statistical mechanics, in two important ways:

Professor Wang, early 1980s in StonyBrook.

(1) I was deeply impressed by Gibbs. I still remember vividly today reading his papers on the phase rule in an obscure journal published in Connecticut. [It is amazing that such an obscure journal was in the library of LianDa.] More important, his little book *Elementary Principles of Statistical Mechanics* converted me to an ardent admirer of his. I was to write in 1963[3]:

> *The beauty of his Elementary Principles of Statistical Mechanics is sheer poetry.*

(2) Statistical mechanics became one of my two major areas of research work. I still remember today Professor Wang excitedly telling me, one day in 1945, the breakthrough Onsager had made in the Ising model. I tried to understand this breakthrough, first in Kunming, later in Chicago in 1947, both without success. But finally in 1949, in a station wagon ride, I learned from Luttinger of a new paper by Kaufman and Onsager. And that led to my lifelong interest in statistical mechanics.

---

[3]C.N. Yang. Selected Papers 1945–1980 With Commentary. Freeman and Company (1983), p. 71.

1948 thesis for my PhD degree

During the first 9 month of 1946 I worked closely with Professor Teller. He had around 6 or 7 graduate students, and met us every week or two for lunch, to discuss our research. He also asked me to grade exercise papers of his students. So I had ample opportunity to observe his style of doing physics. He had very good physical intuition, especially about symmetries in atomic, molecular and nuclear physics. But he lacked patience to fill in the logical steps behind his intuition. For example as early as 1941 he had made, in a paper with Critchfield, statements about complexities in nuclear reactions involving particles with spin, but did not give any proofs. I began to think about how to supply complete proofs.

With Professor Teller,
1982 in Brookhaven National Laboratory.

In the late 1940s low energy nuclear physics was one of the most active fields. In particular there began experiments about correlations, such as β–γ and γ–γ correlations. Theoretical calculations were published about such correlations, showing *very surprisingly* that the final results often were very simple after *unexpected* cancellations. I did some of these calculations and soon realized that the cancellations *must be mathematical consequences of the spherical symmetry of nuclear physics.* But to substantiate such a statement required detailed mathematical analysis. This I succeeded in doing after a few weeks of analysis, and that was how my PhD thesis came about.

With Professor Teller, 1990s.

This thesis greatly increased my appreciation of the power of symmetry considerations in understanding natural laws. It happened that to study the newly discovered "strange particles" it was necessary to first determine their spin, parity and other quantum umbers, i.e. their symmetry properties. Thus I was able, one year after my thesis, to publish a paper on the spin and parity of the π^0 meson. This paper made me famous because it was in direct competition with L. Landau.

I should mention here that this π^0 paper used heavily *field theory*, which I had learned, very thoroughly in 1943–1945, from Professor S. T. Ma (马仕俊).

In Chicago I was interested in using symmetry considerations not only on experiment related problems, such as those in my PhD thesis, but also on a more fundamental problem: the basic equations governing interactions between particles. Thus in 1947 I tried to generalize Weyl's gauge symmetry to non-Abelian groups. This effort met with smooth sailing at the beginning, but soon got into messy technical problems, and I had to give up. Fortunately I did return to it in 1954, at Brookhaven with Robert Mills. We succeeded in overcoming the technical difficulty and published a short paper on it[4]. That paper has now become one of the most important papers in physics after WWII.

With Professors Wu and Ma, 1949.

---

[4]C.N. Yang. Selected Papers 1945–1980 With Commentary. Freeman and Company (1983), p. 19.

Group Theory and the Vibration of Polyatomic Molecules

1942 thesis for BSc degree

GROUP THEORY AND THE VIBRATION OF POLYATOMIC MOLECULES

Cheng-Ning Yang (楊振寧)

INTRODUCTION

Informations about the structure of molecules can always be drawn from the analysis of their vibrational spectra, but owing to the mathematical difficulties involved in the theoretical calculation, only very simple types of molecules can be studied. The method developed by Bethe[1] in 1929, and then more completely by Wigner[2], however, removed considerably this difficulty. It is our purpose here to present the method together with some of the developments after them. A new method of finding the symmetrical coordinates is given (§4), in which the symmetry is preserved from step to step in spite of the existence of redundant coordinates. The theorem in §5 which renders the calculation of the degree of degeneracy very simple is also believed to be new.

The Symmetry of a Molecule

§1 MATHEMATICAL EXPRESSION OF SYMMETRY

There are reasons to suppose that the nuclei in a molecule arrange themselves in symmetrical positions when in equilibrium; i.e. some operations (Consisting of reflections and rotations) bring the molecule into itself. (For molecules containing isotopes this statement must be slightly modified. cf. §18) If we choose a set of rectangular coordinate axes with the origin at the centre of mass of the molecule in equilibrium, each covering operation C can be represented by an orthogonal matrix Γ_c (order: 3×3) so that the point $\begin{pmatrix} X \\ Y \\ Z \end{pmatrix}$ is brought to $\Gamma_c \begin{pmatrix} X \\ Y \\ Z \end{pmatrix}$ by the operation. Let $\boldsymbol{r}_1, \boldsymbol{r}_2, \cdots, \boldsymbol{r}_n$ be a set of coordinates specifying the relative positions of the nuclei (e.g. the distances between the nuclei and the angles between the bonds) in the molecule. When the nuclei vibrate about their positions of equilibrium, these $\boldsymbol{z}$'s vary (cf. §6). Let $R_1, R_2, \cdots,$ R_n be their increments. Further, let $x_1, y_1, z_1, x_2, \cdots, x_N, y_N, z_N$ be the increments of the rectangular coordinates of the N nuclei. For small vibrations the R's are linear in the x's, y's and z's:

$$R = \begin{pmatrix} R_1 \\ R_2 \\ \vdots \\ R_n \end{pmatrix} = B\mathscr{C}, \qquad \text{where } \mathscr{C} = \begin{pmatrix} x_1 \\ y_1 \\ \vdots \\ z_N \end{pmatrix} \tag{1}$$

B being a constant matrix of n rows and $3N$ colomns. Now after the operation C, the molecule is indistinguishable from its original self, and we have a new equation obtained by writing (1) down for the new molecule:

$$\overset{c}{R} = B\overset{c}{\mathscr{C}} \tag{2}$$

Here $\overset{c}{R}$'s are the coordinates of the molecule which will be brought into coincidence with R's by the operation C, and

$$\overset{c}{\mathscr{C}} = \begin{pmatrix} \Gamma_c \begin{pmatrix} x_{C^{-1}1} \\ y_{C^{-1}1} \\ z_{C^{-1}1} \end{pmatrix} \\ \vdots \\ \Gamma_c \begin{pmatrix} x_{C^{-1}N} \\ y_{C^{-1}N} \\ z_{C^{-1}N} \end{pmatrix} \end{pmatrix} \tag{3}$$

where $C^{-1}i$ is the nucleus which will become after the operation C the nucleus i. Let Z_c be a square matrix of order $3N$ with the elements

$$_{ix}(Z_c)_{jx} = \delta_{j,C^{-1}i}, \quad _{ix}(Z_c)_{jy} = 0 \quad \text{etc.} \quad i,j = 1,2,\cdots,N,$$

and let P_c stand for $\begin{pmatrix} \Gamma_c & & & \\ & \Gamma_c & & \\ & & \ddots & \\ & & & \Gamma_c \end{pmatrix}$, then (2) and (3) give

$$\overset{c}{R} = BP_c Z_c \mathscr{C}. \tag{4}$$

This equation holds for every operation C and is the mathematical expression of the symmetry of the molecule.

Group Theory and the Vibration of Polyatomic Molecules

EXAMPLE Consider three equivalent nuclei forming an equilateral triangle. Let C be the operation: Rotation counterclockwise through $120°$ about O. Then

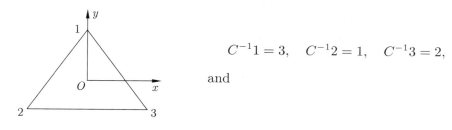

$$C^{-1}1 = 3, \quad C^{-1}2 = 1, \quad C^{-1}3 = 2,$$

and

$$Z_c = \begin{pmatrix} & & & & 1 & 0 & & & \\ 0 & & 0 & & & 1 & & & \\ & & & & 0 & 1 & & & \\ 1 & 0 & & & & & & & \\ & 1 & & & 0 & & & 0 & \\ 0 & 1 & & & & & & & \\ & & & 1 & 0 & & & & \\ & & 0 & & 1 & & & 0 & \\ & & & 0 & 1 & & & & \end{pmatrix}, \quad \varGamma_c = \begin{pmatrix} -\frac{1}{2} & -\frac{\sqrt{3}}{2} & 0 \\ \frac{\sqrt{3}}{2} & -\frac{1}{2} & 0 \\ 0 & 0 & 1 \end{pmatrix}.$$

Let R_1 be the increment of the distance $\overline{12}$.

$$R_1 = \frac{1}{2}x_1 + \frac{\sqrt{3}}{2}y_1 - \frac{1}{2}x_2 - \frac{\sqrt{3}}{2}y_2.$$

Then $\overset{c}{R_1}$ is that of $\overline{31}$, and

$$\overset{c}{R_1} = -\frac{1}{2}x_1 + \frac{\sqrt{3}}{2}y_1 + \frac{1}{2}x_3 - \frac{\sqrt{3}}{2}y_3.$$

Thus

$$B = \begin{pmatrix} \frac{1}{2} & \frac{\sqrt{3}}{2} & 0 & -\frac{1}{2} & -\frac{\sqrt{3}}{2} & 0 & 0 & 0 & 0 \end{pmatrix}.$$

(4) becomes the identity

$$-\frac{1}{2}x_1 + \frac{\sqrt{3}}{2}y_1 + \frac{1}{2}x_3 - \frac{\sqrt{3}}{2}y_3 = \begin{pmatrix} \frac{1}{2} & \frac{\sqrt{3}}{2} & 0 & -\frac{1}{2} & -\frac{\sqrt{3}}{2} & 0 & 0 & 0 & 0 \end{pmatrix} \begin{pmatrix} \varGamma_c & 0 & 0 \\ 0 & \varGamma_c & 0 \\ 0 & 0 & \varGamma_c \end{pmatrix} Z_c \begin{pmatrix} x_1 \\ y_1 \\ \vdots \\ z_3 \end{pmatrix}.$$

§2 FUNDAMENTAL RELATIONSHIP

In some instances the coordinates $R_1, R_2, \cdots, R_n$ are sufficient to determine $\overset{c}{R}_1, \overset{c}{R}_2, \cdots, \overset{c}{R}_n$ for all covering operations C. This is the case if (i) the R's contain only complete sets of equivalent coordinates (e.g. in the last example when $R_2 = \begin{matrix} R_1 \\ \text{increment} \\ R_3 \end{matrix}$ of $1\widehat{\begin{matrix}3\\2\\2\end{matrix}}\begin{matrix}1\\3\\1\end{matrix}$); or if (ii) the R's are all that are necessary to describe the internal structure of the molecule. In both cases we have for small vibrations $\overset{c}{R} = A_c R$, where A_c is in case (i) an orthogonal matrix having as elements 0 or 1, and in case (ii) a matrix of order $n \times n$. By (4),

$$BP_c Z_c \mathscr{C} = \overset{c}{R} = A_c R = A_c B \mathscr{C}.$$

But $\mathscr{C}$ is arbitrary (cf. §6), hence

$$BP_c Z_c = A_c B. \tag{5}$$

This is the fundamental relationship on which all the following deductions are based.

§3 GROUP PROPERTIES

To make further developments we notice that the covering operations C form a group and that the P_c's, Z_c's and A_c's each form a group isomorphic[*] with it. The group is known as the "point group". They are of such importance that their properties have been investigated in detail.[3]

Choice of Internal Coordinates

§4 INDEPENDENT REDUCED COORDINATES

We first choose the coordinates $R_1, R_2, \cdots, R_n$ so that they contain only complete sets of equivalent internal coordinates, and such that they are more than necessary for the determination of the structure of the molecule. The simplest way is to choose the increments of the internuclear distances and the bond angles as the R's. In the example

---

[*]Let $C'C$ be the resultant operation of first operating C and then C', we have

$$P_{c'c} = P_{c'} P_c, \quad Z_{c'c} = Z_{c'} Z_c, \quad \text{but } A_{c'c} = A_c A_{c'}.$$

of §1 we may take the increments of the bonds $\overline{12}$, $\overline{23}$ and $\overline{31}$ as R_1, R_2 and R_3; or those of the lengths $\overline{O1}$, $\overline{O2}$, $\overline{O3}$ and the angles $1\widehat{O}2$, $2\widehat{O}3$ and $3\widehat{O}1$ as the R's. The matrix B can now be determined (§§11, 12). Evidently our choice belongs to the case (i) of §2, so that the A_c's are orthogonal and have as elements 0 or 1. It is plain that $_i(A_c)_j = 0$ if R_i and R_j are not equivalent. We shall make use of the following theorem in group theory[5]:

If A_c form a group of orthogonal matrices, and $W_c^\alpha (\alpha = 1, 2, \cdots, k)$ are the irreducible orthogonal representations of the group, there exists an orthogonal matrix M such that $W_c = M A_c M^{-1}$ is of the form

$$\begin{pmatrix} W_c^1 & & & & & & 0 \\ & W_c^1 & & & & & \\ & & \ddots & & & & \\ & & & W_c^1 & & & \\ & & & & W_c^2 & & \\ & & & & & \ddots & \\ 0 & & & & & & W_c^k \end{pmatrix}. \quad (6)$$

We define* $Q = \begin{pmatrix} Q_1 \\ Q_2 \\ \vdots \\ Q_n \end{pmatrix} = MR$ as the "reduced coordinates"[4]. Evidently

$$\overset{c}{Q} = M\overset{c}{R} = MA_c R = MA_c M^{-1} Q = W_c Q.$$

Now not all the Q's are independent. To select out the independent ones we need the following theorem:

<u>THEOREM</u> It is always possible to drop out some of the Q's so that

(i) the remaining ones are all independent,

(ii) the dropped ones depend on the remaining ones,

and (iii) the remaining ones belong to complete blocks of the group of matrices W_c.

Because of the properties (i) and (ii), the remaining coordinates $\mathscr{R} = \begin{pmatrix} \mathscr{R}_1 \\ \mathscr{R}_2 \\ \vdots \\ \mathscr{R}_{3N-6} \end{pmatrix}$ are

called the "independent reduced coordinates", and from (iii), $\overset{c}{\mathscr{R}} = \mathfrak{A}_c \mathscr{R}$ where $\mathfrak{A}_c$ is

---

*We assume here that Q is real. There is probably no difficulty in handling complex coordinates, but they are not necessary unless molecules of symmetry C_n, C_{nh} or S_n ($n > 2$) are studied.

obtained from W_c by dropping some blocks. These coordinates may also be obtained in other ways, and have been given various names by different authors. They are the "internal symmetry coordinates" of Wilson[7] and the "geometrical symmetry coordinates" of Rosenthal and Murphy[8]. The relation between $\mathscr{R}$ and the "symmetry coordinates" of Howard and Wilson[11,9] will be given in §14.

PROOF OF THE THEOREM Since Q is not entirely arbitrary, there exists a matrix H of order $\nu \times n$ so that

$$HH' = I, \quad HQ = 0, \tag{7}$$

where I is the unit matrix and the $'$ stands for "transposed". We may prove that the rank of H is ν and that there exists a matrix H_1 of order $(n-\nu) \times n$ such that $\begin{pmatrix} H \\ H_1 \end{pmatrix}$ is orthogonal. Put $\begin{pmatrix} H \\ H_1 \end{pmatrix} Q = \begin{pmatrix} 0 \\ Q_1 \end{pmatrix}$, then

$$Q = (H' \ H_1') \begin{pmatrix} 0 \\ Q_1 \end{pmatrix} = H_1' Q_1. \tag{8}$$

We may suppose that $\nu = n - (3N-6)$ so that Q_1 is arbitrary. The equ. $HQ = 0$ must be invariant under an operation C:

$$0 = H\overset{c}{Q} = HW_c Q = HW_c H_1' Q_1.$$

Hence

$$HW_c H_1' = 0.$$

Thus

$$\begin{pmatrix} H \\ H_1 \end{pmatrix} W_c (H' \ H_1') = \begin{pmatrix} HW_c H' & 0 \\ H_1 W_c H' & H_1 W_c H_1' \end{pmatrix}.$$

This matrix is orthogonal (for $\begin{pmatrix} H \\ H_1 \end{pmatrix}$ and W_c are both orthogonal). Hence

$$\begin{pmatrix} H \\ H_1 \end{pmatrix} W_c (H' \ H_1') = \begin{pmatrix} HW_c H' & 0 \\ 0 & H_1 W_c H_1' \end{pmatrix}. \tag{9}$$

Thus $HW_c H'$ is orthogonal and forms a group isomorphic with the one formed by C; i.e.

$$(HW_{c'} H')(HW_c H') = HW_{cc'} H'.$$

By the theorem quoted above there exists an orthogonal matrix J such that $JHW_cH'J' = \omega_c$ is of the form (6); i.e. if the "untermatrices" of W_c are

$$_{\alpha\beta}(W_c)_{\alpha'\beta'} = \delta_{\alpha\alpha'}\delta_{\beta\beta'}W_c^\alpha, \quad \alpha = 1, 2, \cdots, k, \quad \beta = 1, 2, \cdots, n_\alpha,$$

those of ω_c must be

$$_{ab}(\omega_c)_{a'b'} = \delta_{aa'}\delta_{bb'}W_c^a, \quad a = 1, 2, \cdots, k, \quad b = 1, 2, \cdots, \mu_a.$$

By (9), the irreducible blocks of ω_c must all be that of W_c, hence $\mu_a \leq n_\alpha$. Now

$$(JH)W_c = JHW_c(H' \ H_1')\begin{pmatrix} H \\ H_1 \end{pmatrix} = (JHW_cH' \ 0)\begin{pmatrix} H \\ H_1 \end{pmatrix} = \omega_c JH \qquad (10)$$

i.e.

$$_{ab}(JH)_{\alpha\beta}\, W_c^\alpha = W_c^a \,_{ab}(JH)_{\alpha\beta}.$$

But W_c^α is irreducible, hence[6]

$$_{ab}(JH)_{\alpha\beta} = \delta_{a\alpha}\,_b\lambda_\beta^\alpha I, \qquad (11)$$

where $_b\lambda_\beta^\alpha$ is an ordinary number. Since JH is of rank ν, the submatrix $\left(_b\lambda_\beta^\alpha\right)$ ($b = 1, 2, 3, \cdots, \mu_\alpha$; $\beta = 1, 2, 3, \cdots, n_\alpha$) is of rank μ_α. Thus for every α there exists a set K_α of μ_α integers all $\leq n_\alpha$, such that the square matrix $\left(_b\lambda_\beta^\alpha\right)$ ($b = 1, 2, \cdots, \mu_\alpha$; β in K_α) is nonsingular. Hence JH may be divided into two untermatrices, one (of order $\nu \times \nu$) composed of those untermatrices (11) for which β is in the set k_α, and the other those for which β is not in K_α. The former is evidently nonsingular. Let H_2 and H_3 be the corresponding untermatrices of $H = J^{-1}(JH)$. Evidently a transposition of columns may bring H into the form $(H_2 \ H_3)$. Now the constraint on Q is $HQ = 0$. Hence a corresponding transposition of the rows of Q bring it to $\begin{pmatrix} Q_2 \\ Q_3 \end{pmatrix}$ so that

$$H_2Q_2 + H_3Q_3 = 0.$$

Since H_2 is nonsingular, Q_3 may be chosen as the independent variable, Q_2 being dependent on it. Thus the latter can be dropped and the conditions (i), (ii) and (iii) are satisfied.

§5 CALCULATION OF SPUR ($\mathfrak{A}_c$)

Suppose that the constraints on the R's are given by

$$\mathscr{D}R = 0 \quad \text{where} \quad \mathscr{D}\mathscr{D}' = I. \qquad (12)$$

THEOREM

$$\mathrm{Spur}(\mathfrak{A}_c) = \mathrm{Spur}(A_c)' - \mathrm{Spur}(\mathscr{D}A_c\mathscr{D}') \tag{13}$$

This theorem makes the calculation of $\mathrm{Spur}(\mathfrak{A}_c)$ very simple (because the elements of A_c are 0 or 1), and before the transformation from R the Q is carried out. Also it enables us to calculate the contributions of the different constraints separately.

PROOF   The constraints on $R_1, R_2, \cdots, R_n$ are $\mathscr{D}R = 0$. Hence we may take the matrix H of (7) to be $\mathscr{D}M'$. With the notation used there we have

$$\overset{c}{Q}_1 = H_1 \overset{c}{Q} = H_1 W_c Q = H_1 W_c H_1' Q_1. \tag{14}$$

Since the $\mathscr{R}$'s are all independent, there exists M_1 such that $Q_1 = M_1 \mathscr{R}$. Hence

$$\overset{c}{Q}_1 = M_1 \overset{c}{\mathscr{R}} = M_1 \mathfrak{A}_c \mathscr{R} = M_1 \mathfrak{A}_c M_1^{-1} Q_1.$$

On comparison with (14) it follows that $H_1 W_c H_1' = M_1 \mathfrak{A}_c M_1^{-1}$. Thus

$$\mathrm{Spur}(\mathfrak{A}_c) = \mathrm{Spur}(H_1 W_c H_1')$$
$$= \mathrm{Spur}\left[\begin{pmatrix} H \\ H_1 \end{pmatrix} W_c (H' \ H_1')\right] - \mathrm{Spur}(H W_c H')$$
$$= \mathrm{Spur}(A_c) - \mathrm{Spur}(\mathscr{D}A_c\mathscr{D}').$$

EXAMPLE   Consider the molecule $\mathrm{CH_3Cl}$. Take

$R_1, R_2, R_3, R_4 = $ increments of the distances $\overline{\mathrm{CCl}}$, $\overline{\mathrm{CH}_\alpha}$, $\overline{\mathrm{CH}_\beta}$, $\overline{\mathrm{CH}_\gamma}$,

$R_5, R_6, R_7 = $ increments of the angles $\mathrm{Cl}\widehat{\mathrm{C}}\mathrm{H}_\alpha$, $\mathrm{Cl}\widehat{\mathrm{C}}\mathrm{H}_\beta$, $\mathrm{Cl}\widehat{\mathrm{C}}\mathrm{H}_\gamma$,

$R_8, R_9, R_{10} = $ increments of the angles $\mathrm{H}_\beta\widehat{\mathrm{C}}\mathrm{H}_\gamma$, $\mathrm{H}_\gamma\widehat{\mathrm{C}}\mathrm{H}_\alpha$, $\mathrm{H}_\alpha\widehat{\mathrm{C}}\mathrm{H}_\beta$.

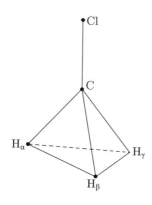

The constraint is

$$[R_5 + R_6 + R_7]f + [R_8 + R_9 + R_{10}] = 0,$$

where f is a constant. Thus

$$\mathscr{D} = \sqrt{\frac{1}{3f^2 + 3}}(0\ 0\ 0\ 0\ f\ f\ f\ 1\ 1\ 1).$$

The theorem leads to

Operation C		Spur(A_c)	Spur($\mathscr{D}A_c\mathscr{D}$)*	Spur($\mathfrak{A}_c$)
No Motion	: C_1	10	1	9
Rotation about $\overline{CCl}$ through	$120°$: C_2	1	1	0
	$240°$: C_3			
Reflection about the plane	$Cl\,CH_\alpha$: C_4	4	1	3
	$Cl\,CH_\beta$: C_5			
	$Cl\,CH_\gamma$: C_6			

The Kinetic and the Potential Energies

§6 CHOICE OF AXES

Suppose that $\begin{pmatrix} R_1 \\ R_2 \\ \vdots \\ R_{3N-6} \end{pmatrix} = R$ are the increments of $3N - 6$ independent internal coordinates. For small vibrations, $R = B\mathscr{C}$, where $\mathscr{C} = \begin{pmatrix} x_1 \\ \vdots \\ z_N \end{pmatrix}$ is defined in §1.

Suppose that the equilibrium positions of the nuclei are, in rectangular coordinates, $(X_1, Y_1, Z_1), (X_2, Y_2, Z_2), \cdots, (X_N, Y_N, Z_N)$. Let $m_1, m_2, \cdots, m_N$ be their masses. Write

$$F = \begin{pmatrix} m_1 & 0 & 0 & m_2 & 0 & 0 & \cdots & 0 \\ 0 & m_1 & 0 & 0 & m_2 & 0 & \cdots & 0 \\ 0 & 0 & m_1 & 0 & 0 & m_2 & \cdots & m_N \\ 0 & m_1 Z_1 & -m_1 Y_1 & 0 & m_2 Z_2 & -m_2 Y_2 & \cdots & -m_N Y_N \\ -m_1 Z_1 & 0 & m_1 X_1 & -m_2 Z_2 & 0 & m_2 X_2 & \cdots & m_N X_N \\ m_1 Y_1 & -m_1 X_1 & 0 & m_2 Y_2 & -m_2 X_2 & 0 & \cdots & 0 \end{pmatrix}$$

---

* From this it is immediately seen that any constraint having the same coefficient for equivalent coordinates contributes 1 to Spur ($\mathscr{D}A_c\mathscr{D}'$).

Then the first column of $B \begin{pmatrix} \frac{1}{m_1} & & \\ & \ddots & \\ & & \frac{1}{m_N} \end{pmatrix} F'$ is the matrix $B \begin{pmatrix} 1 \\ 0 \\ 0 \\ 1 \\ 0 \\ 0 \\ \vdots \\ 0 \end{pmatrix}$ which is the value of R when all the nuclei are displaced by 1 unit of length along the X-axis. But the R's are the increments of internal variables, hence $B \begin{pmatrix} 1 \\ 0 \\ 0 \\ 1 \\ 0 \\ 0 \\ \vdots \\ 0 \end{pmatrix} = 0$. In the same way we can show that

$$B \begin{pmatrix} \frac{1}{m_1} & & \\ & \ddots & \\ & & \frac{1}{m_N} \end{pmatrix} F' = 0.$$

We have thus far described the the molecule in a certain system of rectangular coordinates. But when the molecule moves and rotates in space as well as vibrates, there remains an arbitrariness in fixing the coordinate axes to the molecule. The following method of choosing these axes is, however, the most preferable. The $3N - 6$ internal variables $R_1, R_2, \cdots, R_{3N-6}$ is determined uniquely from the structure of the molecule. We define $x_1, y_1, \cdots, z_N$ by

$$\mathscr{C} = \begin{pmatrix} x_1 \\ \vdots \\ z_N \end{pmatrix} = \begin{pmatrix} B \\ F \end{pmatrix}^{-1} \begin{pmatrix} R \\ 0 \end{pmatrix}, \tag{15}$$

so that

$$F\mathscr{C} = 0. \tag{16}$$

From the definition of B, we conclude that a system of axes of reference can be found so that the positions of the nuclei are $(X_1+x_1, Y_1+y_1, Z_1+z_1)$, $(X_2+x_2, Y_2+y_2, Z_2+z_2)$, $\cdots$, $(X_N+x_N, Y_N+y_N, Z_N+z_N)$ when the molecule is not very much distorted from its equilibrium structure. We have thus 6 external variables specifying the position and orientation of the axes in space and $3N-6$ variables $R_1, R_2, \cdots, R_{3N-6}$ specifying by means of (15) the positions of the nuclei with respect to these axes.

§7 THE KINETIC ENERGY IN TERMS OF R

Equ. (16) expresses the facts that the origin of our moving axes is at any time the centre of mass of the molecule, and that the moment of momentum of the molecule in this system of reference is of the order of $m x \dot{x}$. Thus if the molecule rotates in space with an angular velocity ω about its centre of mass, its kinetic energy is

$$T_{\text{centre of mass}} + T_{\text{rotation}} + \frac{1}{2}\sum_i m_i(\dot{x}_i^2 + \dot{y}_i^2 + \dot{z}_i^2) + \text{term} \sim m x \dot{x} \omega.$$

But

$$T_{\text{rotation}} = (\text{K.E. of rotation if } \mathscr{C}=0) + \text{term} \sim m\omega^2 x X.$$

Now $T_{\text{centre of mass}} + $ (K.E. of rotation if $\mathscr{C}=0$) depends on the external variables only (together, of course, with their time derivatives). And in a gas, due to thermal agitation, $\omega X \sim \dot{x}$.

Thus in the first approximation the equ. of motion is to be derived from (cf. §§15, 16)

$$T = \frac{1}{2}\sum_i m_i(\dot{x}_i^2 + \dot{y}_i^2 + \dot{z}_i^2) = \frac{1}{2}(\dot{R}'\ 0)\begin{pmatrix}B\\F\end{pmatrix}^{-1\prime}\begin{pmatrix}m_1 & & \\ & \ddots & \\ & & m_N\end{pmatrix}\begin{pmatrix}B\\F\end{pmatrix}^{-1}\begin{pmatrix}\dot{R}\\0\end{pmatrix}.$$

But

$$\begin{pmatrix}B\\F\end{pmatrix}\begin{pmatrix}\frac{1}{m_1} & & \\ & \ddots & \\ & & \frac{1}{m_N}\end{pmatrix}\begin{pmatrix}B' & F'\end{pmatrix} = \begin{pmatrix} B\begin{pmatrix}\frac{1}{m_1} & & \\ & \ddots & \\ & & \frac{1}{m_N}\end{pmatrix}B' & 0 \\ 0 & F\begin{pmatrix}\frac{1}{m_1} & & \\ & \ddots & \\ & & \frac{1}{m_N}\end{pmatrix}F' \end{pmatrix}.$$

Hence writing

$$G = B\begin{pmatrix}\frac{1}{m_1} & & \\ & \ddots & \\ & & \frac{1}{m_N}\end{pmatrix}B', \qquad (17)$$

we have
$$2T = \dot{R}'G^{-1}\dot{R}. \qquad (18)$$

§8 THE KINETIC ENERGY IN TERMS OF $\mathscr{R}$

If we take the independent reduced coordinates $\mathscr{R}$ (§4) to be the R's of the last section, the results may be summarized:

$$\mathscr{R} = \mathscr{L}\mathscr{C}, \qquad \mathscr{G} = \mathscr{L}\begin{pmatrix} \frac{1}{m_1} & & \\ & \ddots & \\ & & \frac{1}{m_N} \end{pmatrix}\mathscr{L}', \qquad 2T = \dot{\mathscr{R}}'\mathscr{G}^{-1}\dot{\mathscr{R}}. \qquad (19)$$

Now by (5),
$$\mathscr{L}P_c Z_c = \mathfrak{A}_c \mathscr{L} \qquad (\mathfrak{A}_c \text{ stands here for } A_c, \text{ cf. §4})$$

so that
$$\mathscr{G} = \mathscr{L}\begin{pmatrix} \frac{1}{m_1} & & \\ & \ddots & \\ & & \frac{1}{m_N} \end{pmatrix}\mathscr{L}' = \mathfrak{A}'_c \mathscr{L} P_c Z_c \begin{pmatrix} \frac{1}{m_1} & & \\ & \ddots & \\ & & \frac{1}{m_N} \end{pmatrix} Z'_c P'_c \mathscr{L}' \mathfrak{A}_c.$$

But $\begin{pmatrix} \frac{1}{m_1} & & \\ & \ddots & \\ & & \frac{1}{m_N} \end{pmatrix}$, P_c and Z_c commute[*] with each other, hence

$$\mathscr{G} = \mathfrak{A}'_c \mathscr{L}\begin{pmatrix} \frac{1}{m_1} & & \\ & \ddots & \\ & & \frac{1}{m_N} \end{pmatrix}\mathscr{L}' \mathfrak{A}_c = \mathfrak{A}'_c \mathscr{G} \mathfrak{A}_c.$$

Suppose
$$_{\alpha\beta}(\mathfrak{A}_c)_{\alpha'\beta'} = \delta_{\alpha\alpha'}\delta_{\beta\beta'}W_c^\alpha, \quad \alpha = 1,2,\cdots,k, \ \beta = 1,2,\cdots,n_\alpha, \qquad (20)$$

where W_c^α is irreducible. (cf. §4) Since $\mathfrak{A}_c \mathscr{G} = \mathscr{G}\mathfrak{A}_c$, we have

$$W_c^\alpha \ _{\alpha\beta}(\mathscr{G})_{\alpha'\beta'} = \ _{\alpha\beta}(\mathscr{G})_{\alpha'\beta'} W_c^{\alpha'}.$$

Hence[6]
$$_{\alpha\beta}(\mathscr{G})_{\alpha'\beta'} = \delta_{\alpha\alpha'} \ _\beta g^\alpha_{\beta'} I. \qquad (21)$$

If W_c^α is of order d_α, this shows that the nonvanishing elements of $\mathscr{G}$ are in $\sum_{\alpha=1}^{k} d_\alpha$ diagonal blocks of which d_α are identical of dimension $n_\alpha \times n_\alpha$.

---

[*]cf. §18.

§9 THE POTENTIAL ENERGY IN TERMS OF $\mathscr{R}$

The potential energy depends on the internal coordinates only. For small vibrations, it is approximately equal to $\frac{1}{2}\mathscr{R}'\mathfrak{V}\mathscr{R}$, where $\mathfrak{V}$ is a positive symmetrical matrix, because when in equilibrium the molecule has a minimum potential energy. The covering operations leave the potential energy unchanged:

$$\frac{1}{2}\mathscr{R}'\mathfrak{V}\mathscr{R} = \frac{1}{2}\overset{c}{\mathscr{R}'}\mathfrak{V}\overset{c}{\mathscr{R}} = \frac{1}{2}\mathscr{R}'\mathfrak{A}_c\mathscr{R}.$$

But $\mathscr{R}$ is arbitrary, hence

$$\mathfrak{V} = \mathfrak{A}'_c\mathfrak{V}\mathfrak{A}_c.$$

Thus we have, just as we had (21),

$$_{\alpha\beta}(\mathfrak{V})_{\alpha'\beta'} = \delta_{\alpha\alpha'}\,_{\beta}v_{\beta'}{}^{\alpha}I, \quad \alpha = 1,2,\cdots,k, \quad \beta = 1,2,\cdots,n_\alpha. \tag{22}$$

There are therefore totally $\sum_{\alpha=1}^{k} n_\alpha^2$ independent constants in the potential energy, which are usually unknowns.[11]

Calculation of the Kinetic Energy

§10 THE TRANSFORMATION MATRIX M [7]

In §4 the orthogonal matrix M was introduced to reduce A_c. Since

$$A_c = \begin{pmatrix} A_{c1} & & 0 \\ & A_{c2} & \\ 0 & & \ddots \end{pmatrix}$$

where each A_{cp} belongs to a set of equivalent coordinates, (§4)

$$M = \begin{pmatrix} M_1 & & 0 \\ & M_2 & \\ 0 & & \ddots \end{pmatrix}$$

where M_p is orthogonal. Hence $W_{cp} = M_p A_{cp} M_p^{-1}$ is of the form (6), i.e.

$$W_{cp} = \begin{pmatrix} W_{cp}^1 & & & \\ & W_{cp}^1 & & \\ & & \ddots & \\ & & & W_{cp}^k \end{pmatrix}.$$

Suppose that W_{cp}^1 is of dimension $d \times d$. Denote the first d rows of M_p by

$$L = \begin{pmatrix} {}_1(M_p)_1 & {}_1(M_p)_2 & \cdots & \cdots \\ \cdots & \cdots & \cdots & \cdots \\ {}_d(M_p)_1 & \cdots & \cdots & {}_d(M_p)_{g_p} \end{pmatrix},$$

where $g_p =$ no. of equivalent coordinates R of the p-th set.

Now
$$W_{cp} M_p = M_p A_{cp}.$$

Hence
$$W_{cp}^1 L = L A_{cp}.$$

Thus
$$L'L = A'_{cp} L'L A_{cp}, \quad \text{i.e.} \quad (L'L) A_{cp} = A_{cp}(L'L).$$

Hence
$$\sum_{t=1}^{g_p} {}_1(L'L)_t \, {}_t(A_{cp})_2 = \sum_{t=1}^{g_p} {}_1(A_{cp})_t \, {}_t(L'L)_2.$$

There exists an operation C which brings the second coordinate of the p-th set to the first, for which $\overset{c}{R_1} = R_2$, so that ${}_1(A_{cp})_t = \delta_{t2}$, ${}_t(A_{cp})_2 = \delta_{t1}$. Hence ${}_1(L'L)_1 = {}_2(L'L)_2$. Thus all diagonal elements of $L'L$ are equal, and

$$\sum_{\gamma=1}^{d} {}_\gamma(M_p)_t^2 = {}_t(L'L)_t = \frac{1}{g_p} \text{Spur}(L'L) = \frac{1}{g_p} \text{Spur}(LL') = \frac{d}{g_p}.$$

Now d can be determined from the values of Spur $(\mathfrak{A}_c)$, (cf. equ. (39)) so that this relation facilitates the calculation of M.

§11 VECTORIAL NOTATION [7]

The matrix $\mathscr{G} = \mathscr{L} \begin{pmatrix} \frac{1}{m_1} & & \\ & \ddots & \\ & & \frac{1}{m_N} \end{pmatrix} \mathscr{L}'$ is calculated by first computing $\mathscr{L}$. Now $\mathscr{R}$ is a submatrix of Q. Hence

$$\mathscr{L} = M_1 B, \quad \text{where } M_1 = {}_{\text{some rows}}(M)_{\text{all columns}}. \tag{23}$$

Thus we have to find B first.
Now
$$R = B\mathscr{C}.$$

Write
$$(x_t \quad y_t \quad z_t) = \vec{\mathscr{C}_t}, \quad ({}_k B_{tx} \quad {}_k B_{ty} \quad {}_k B_{tz}) = \vec{{}_k S_t}.$$

Then $R_k = \sum_t {}_k\vec{S_t} \cdot \vec{\mathscr{C}_t}$. In a similar way we shall write

$$_{\alpha\beta\gamma}(\mathscr{L})_t = \sum_{k=1}^n {}_{\alpha\beta\gamma}(M_1)_k \, {}_k\vec{S_t} = {}_{\alpha\beta\gamma}\vec{S_t}, \quad \gamma = 1, 2, \cdots, d_\alpha, \quad t = 1, 2, \cdots, N. \quad \text{(cf. (21))}$$
(24)

§12 EXPLICIT EXPRESSION OF ${}_k\vec{S_t}$ [7]

The use of the vectors $\vec{S}$ is advantageous because they are independent of the coordinate axes. Two kinds of R's are commonly used:

(i) $R =$ the increment of the bond between the nuclei t' and t'' in length. Evidently when $\vec{\mathscr{C}_{t'}} = \vec{\mathscr{C}_{t''}} = 0$, $R = 0$. Hence

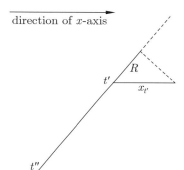

$\vec{S_t} = 0$ for all t except t' or t''.

If $0 = y_{t'} = z_{t'} = x_{t''} = y_{t''} = z_{t''}$, we have $R = x_{t'} \cos\theta$, so that $B_{t'x} = \cos\theta$. Hence

$$\begin{cases} \vec{S_{t'}} = \text{unit vector from } t'' \text{ to } t'. \\ \vec{S_{t''}} = -\vec{S_{t'}}. \\ \text{all other } \vec{S}_t \text{ are 0.} \end{cases} \quad (25)$$

(ii) $R =$ the increment of the angle $t' - t - t''$. Evidently all $\vec{S}$ are 0 except $\vec{S_t}, \vec{S_{t'}}$ and $\vec{S_{t''}}$. Suppose that

$$\vec{S_{t'}} = x'\vec{\xi'} + y'\vec{\eta'} + z'\vec{\zeta'}.$$

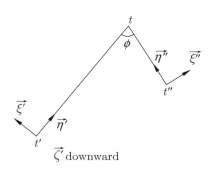

When $\vec{\mathscr{C}_t} = \vec{\mathscr{C}_{t''}} = 0$, $\vec{\mathscr{C}_{t'}} = \vec{\xi'}$, we have, since t' alone is displaced, $R = \frac{1}{\epsilon}$. Hence $x' = \frac{1}{\epsilon}$. Similarly by considering the case when $\vec{\mathscr{C}_t} = \vec{\mathscr{C}_{t''}} = 0$, $\vec{\mathscr{C}_{t'}} = \vec{\eta'}$, we get $y' = 0$. Proceeding in this way we arrive finally at the first line of (26). Now R vanishes when $\vec{\mathscr{C}_{t'}} = \vec{\mathscr{C}_t} = \vec{\mathscr{C}_{t''}}$. Thus $(\vec{S_{t'}} + \vec{S_{t''}} + \vec{S_t}) = 0$. So that

$$\begin{cases} \vec{S_{t'}} = \frac{1}{\epsilon}\vec{\xi'}, \quad \vec{S_{t''}} = \frac{1}{\gamma}\vec{\xi''} \\ \vec{S_t} = -(\vec{S_{t'}} + \vec{S_{t''}}) \\ \text{all other } \vec{S} = 0 \end{cases} \quad (26)$$

From (25) and (26) we can write down, for all k and t, the vector ${}_k\vec{S_t}$ from which $\vec{S}$ may be calculated through (24).

§13 KINETIC ENERGY IN TERMS OF THE $\vec{\mathscr{S}}$'S [7]

We shall show however, that not all the $\vec{\mathscr{S}}$'s are needed for the calculation of $\mathscr{G}$.
Take (5)

$$\mathscr{L}P_c Z_c = \mathfrak{A}_c \mathscr{L}. \tag{27}$$

By (20)

$$\sum_{t'=1}^{N} {}_{\alpha\beta}\mathscr{L}_{t'} \Gamma_c \delta_{t,c^{-1}t'} = W^\alpha_c{}_{\alpha\beta}\mathscr{L}_t, \quad (\Gamma_c \text{ is defined in §2})$$

i.e.

$${}_{\alpha\beta}\mathscr{L}_{ct}\Gamma_c = W^\alpha_c{}_{\alpha\beta}\mathscr{L}_t.$$

Hence for any β and β',

$$\Gamma'_c{}_{\alpha\beta}\mathscr{L}'_{ct}{}_{\alpha\beta'}\mathscr{L}_{ct}\Gamma_c = {}_{\alpha\beta}\mathscr{L}'_t{}_{\alpha\beta'}\mathscr{L}_t,$$

so that

$$\mathrm{Spur}({}_{\alpha\beta}\mathscr{L}'_{ct}{}_{\alpha\beta'}\mathscr{L}_{ct}) = \mathrm{Spur}({}_{\alpha\beta}\mathscr{L}'_t{}_{\alpha\beta'}\mathscr{L}_t). \tag{28}$$

Now by (21),

$${}_\beta g^\alpha_{\beta'} I = {}_{\alpha\beta}\mathscr{G}_{\alpha\beta'} = \sum_p {}_{\alpha\beta}\mathscr{L}_p \left(\frac{1}{m_p} I\right)_{\alpha\beta'} \mathscr{L}'_p, \tag{29}$$

where each p refer to a set of equivalent nuclei.
From (27)

$${}_{\alpha\beta}(\mathscr{L})_p {}_p(P_c)_p {}_p(Z_c)_p = W^\alpha_c{}_{\alpha\beta}(\mathscr{L})_p.$$

Hence

$${}_{\alpha\beta}(\mathscr{L})_p {}_{\alpha\beta'}(\mathscr{L})'_p = W^\alpha_c{}_{\alpha\beta}\mathscr{L}_p {}_{\alpha\beta'}\mathscr{L}'_p,$$

i.e.

$$({}_{\alpha\beta}\mathscr{L}_p {}_{\alpha\beta'}\mathscr{L}'_p) W^\alpha_c = W^\alpha_c ({}_{\alpha\beta}\mathscr{L}_p {}_{\alpha\beta'}\mathscr{L}'_p).$$

Thus[6]

$${}_{\alpha\beta}\mathscr{L}_p {}_{\alpha\beta'}\mathscr{L}'_p = {}_\beta \ell^{\alpha p}_{\beta'} I, \tag{30}$$

where ${}_\beta \ell^{\alpha p}_{\beta'}$ is an ordinary number. From (29) we have,

$${}_\beta g^\alpha_{\beta'} I = \sum_p \frac{1}{m_p} {}_\beta \ell^{\alpha p}_{\beta'} I. \tag{31}$$

Hence

$${}_\beta \ell^{\alpha p}_{\beta'} d_\alpha = \mathrm{Spur}({}_{\alpha\beta}\mathscr{L}_p {}_{\alpha\beta'}\mathscr{L}'_p) = \mathrm{Spur}({}_{\alpha\beta'}\mathscr{L}'_p {}_{\alpha\beta}\mathscr{L}_p) = \sum_{t \text{ in } p} \mathrm{Spur}({}_{\alpha\beta'}\mathscr{L}'_t {}_{\alpha\beta}\mathscr{L}_t). \tag{32}$$

Now consider the sum of (28). Since ct goes over all values of t in p when c goes over all the operations of the group,

$$\mathrm{Spur}(_{\alpha\beta}\mathscr{L}'_{ct}\,_{\alpha\beta'}\mathscr{L}_{ct}) = \mathrm{Spur}(_{\alpha\beta}\mathscr{L}'_{ct'}\,_{\alpha\beta'}\mathscr{L}_{ct'}),$$

if t and t' are in the same p-th set. Thus (32) becomes

$$_{\beta}\mathscr{C}^{\alpha p}_{\beta'}d_\alpha = g_p\mathrm{Spur}(_{\alpha\beta}\mathscr{L}_t\,_{\alpha\beta'}\mathscr{L}'_t),$$

where t represents any nucleus of the p-th set. Substitute into (31) and make use of (24). We obtain:

$$_{\beta}g^{\alpha}_{\beta'} = \frac{1}{d_\alpha}\sum_p \frac{g_p}{m_p}\sum_{\gamma=1}^{d_\alpha} {}_{\alpha\beta\gamma}\vec{S}_t \cdot {}_{\alpha\beta'\gamma}\vec{S}_{t'}. \tag{33}$$

Thus in calculation $\mathscr{G}$ ($_{\alpha\beta}\mathscr{G}_{\alpha'\beta'} = \delta_{\alpha\alpha'}\,_{\beta}g^{\alpha}_{\beta'}I$) we have only to know $_{\alpha\beta\gamma}\vec{S}_t$ for all α ($= 1, 2, \cdots, k$), β ($= 1, 2, \cdots, n_d$) and γ ($= 1, 2, \cdots, d_\alpha$); but only for <u>one</u> t in each p. From (24) we see that not all $\vec{S}$ are needed.

Secular Equation and Degeneracy

§14   NORMAL COORDINATES

In terms of the independent reduced coordinates $\mathscr{R}$ defined in §4, the kinetic energy and the potential energy are $\frac{1}{2}\dot{\mathscr{R}}'\mathscr{G}^{-1}\dot{\mathscr{R}}$ and $\frac{1}{2}\mathscr{R}'\mathscr{V}\mathscr{R}$, (§§8, 9) where $\mathscr{G}$ is positive definite. Let $\lambda_1, \lambda_2, \cdots, \lambda_{3N-6}$ be the roots of the equ.

$$|\lambda\mathscr{G}^{-1} - \mathscr{V}| = 0, \quad \text{i.e.} \quad |\lambda I - \mathscr{V}\mathscr{G}| = 0. \tag{34}$$

Then there exists a matrix L such that

$$L\mathscr{G}L' = I, \qquad L^{-1\prime}\mathscr{V}L^{-1} = \begin{pmatrix} \lambda_1 & & & 0 \\ & \lambda_2 & & \\ & & \ddots & \\ 0 & & & \lambda_{3N-6} \end{pmatrix} = \Lambda.$$

Put

$$\mathscr{N} = \begin{pmatrix} \mathscr{N}_1 \\ \vdots \\ \mathscr{N}_{3N-6} \end{pmatrix} = L\mathscr{R},$$

we get immediately

$$2\times\text{kinetic energy} = \dot{\mathscr{R}}'\mathscr{G}^{-1}\dot{\mathscr{R}} = \dot{\mathfrak{N}}'\mathfrak{N} = \sum_{i=1}^{3N-6}\dot{\mathfrak{N}}_i^2, \tag{35}$$

$$2\times\text{potential energy} = \mathscr{R}'\mathscr{V}\mathscr{R} = \mathfrak{N}'\Lambda\mathfrak{N} = \sum_i \lambda_i \mathfrak{N}_i^2. \tag{36}$$

These coordinates $\mathfrak{N}_i$ are known as the "normal coordinate" 's. They are obtained by first solving (34), thus getting λ_i; and then determining L from

$$L\mathscr{G}\mathscr{V} = \Lambda L. \tag{37}$$

This L must be normalized by

$$L\mathscr{G}L' = I. \tag{38}$$

The normal coordinates are then calculated from $\mathfrak{N} = L\mathscr{R}$.

Now we shall be able to see the role that symmetry plays in simplifying the calculations. Eqs. (21) and (22) show the secular equ. (34) is factored into $\sum_{\alpha=1}^{k} d_\alpha$ equations of which d_α are identical and are of the n_α-th degree. The labor of solving for the λ's is considerably saved. Moreover, the no. of unknown constants in $\mathscr{V}$ is reduced.

The normal coordinates are also a special form of the "symmetry coordinates" introduced by Howard and Wilson.[21] The most general form of the symmetry coordinates $\mathfrak{M}$ is given by $\mathfrak{N} = \mathscr{L}\mathfrak{M}$ where

$$_{\alpha\beta}\mathscr{L}_{\alpha'\beta'} = \delta_{\alpha\alpha'}\,_\beta\ell^\alpha_{\beta'}U^\alpha, \quad \alpha = 1, 2, \cdots, k, \ \beta = 1, 2, \cdots, n_a,$$

in which U^α is orthogonal and $_\beta\ell^\alpha_{\beta'}$ form an orthogonal matrix when β and β' range over $1, 2, \cdots, n_\alpha$.

§15 SOLUTION OF THE PHYSICAL PROBLEM—CLASSICAL THEORY

The Lagrangian is, from (35) and (36),

$$\frac{1}{2}\sum_{i=1}^{3N-6}(\dot{\mathfrak{N}}_i^2 - \lambda_i \mathfrak{N}_i^2).$$

The equs. of motion are therefore

$$\ddot{\mathfrak{N}}_i + \lambda_i \mathfrak{N}_i = 0, \quad i = 1, 2, \cdots, 3N - 6.$$

Hence

$$\mathfrak{N}_i = (\mathfrak{N}_i)_0 \cos(\sqrt{\lambda_i}t + \phi_i).$$

The frequencies of vibration are thus

$$\frac{\sqrt{\lambda_1}}{2\pi}, \frac{\sqrt{\lambda_2}}{2\pi}, \ldots, \frac{\sqrt{\lambda_{3N-6}}}{2\pi}.$$

This is correct only in the first approximation; but it serves to give almost all our present knowledge about the forces within the molecules.

§16 SOLUTION OF THE PHYSICAL PROBLEM—QUANTUM THEORY

Now

$$T = \frac{1}{2}\sum_i \dot{\mathfrak{N}}_i^2, \quad V = \frac{1}{2}\sum_i \lambda_i \mathfrak{N}_i^2.$$

The wave equation is most easily obtained from the variational formulation of the problem:

$$\delta \int \left\{ \sum \hbar^2 g^{ij} \frac{\partial \psi^*}{\partial \mathfrak{N}_i} \frac{\partial \psi}{\partial \mathfrak{N}_j} + (V - E)\psi^*\psi \right\} \sqrt{g}\, d\mathfrak{N}_1 \cdots d\mathfrak{N}_{3N-6} = 0,$$

where g^{ij} is given by $T = g_{ij}\dot{x}^i \dot{x}^j$. Hence

$$\sum_i \frac{1}{2}\left(-\hbar^2 \frac{\partial^2}{\partial \mathfrak{N}_i^2} + \lambda_i \mathfrak{N}_i^2\right)\psi = E\psi.$$

The electric moment in any direction is, in the first approximation

$$D_0 + \sum_{i=1}^{3N-6} D_i \mathfrak{N}_i$$

where D_i is immediately calculable from (15). Thus the wave function and selection rules, hence the frequencies in the vibrational spectra of the molecule, are the same as those of $3N-6$ independent harmonic oscillators with coordinates $\mathfrak{N}_1, \mathfrak{N}_2, \ldots, \mathfrak{N}_{3N-6}$. The frequencies are therefore exactly those obtained classically.

§17 THE DEGREE OF DEGENERACY

We have seen that the secular equ. (34) is factorized into d_1 identical equs. of the n_1-th degree, d_2 identical equs. of the n_2-th degree, $\cdots$. There must therefore be n_1 frequencies each corresponding to d_1 different normal moods of vibration (d_1-fold degenerate). The nos. $d_1, d_2, \cdots$, are accordingly called "the degrees of degeneracy".

It is interesting that they can be determined together with $n_1, n_2, \cdots$, without carrying out the calculation of the normal coordinates. For, by (20)

$$\sum_\alpha n_\alpha \mathrm{Spur}(W_c^\alpha) = \mathrm{Spur}(\mathfrak{A}_c).$$

But
$$\sum_c \mathrm{Spur}(W_c^\alpha)\mathrm{Spur}(W_c^\beta) = \delta_{\alpha\beta}h,$$

where $h =$ order of the group. Hence

$$n_\alpha = \frac{1}{h}\sum_c \mathrm{Spur}(\mathfrak{A}_c)\mathrm{Spur}(W_c^\alpha). \tag{39}$$

Now both $\mathrm{Spur}(W_c^\alpha)$ and d_α can be found[3] from a table of characters of the point group, and $\mathrm{Spur}(\mathfrak{A}_c)$ can be calculated through the theorem in §5, so that d_α and n_α are easily calculable from (39).

EXAMPLE We consider again the molecule CH_3Cl. (§5) From the left table of characters we get $n_1 = 3$, $n_2 = 0$, $n_3 = 3$. It follows that the secular equ. is factored into 3 cubic equs. 2 of which are identical.

Representation	d_α	$\mathrm{Spur}(W_c)$		
		$c = c_1$	$c = c_2, c_3$	$c = c_4, c_5, c_6$
W^1	1	1	1	1
W^2	1	1	1	-1
W^3	2	2	-1	0

We conclude that there are totally 6 vibrational frequencies, 3 of which are doubly degenerate.

The no. of independent constants in the potential energy is (§9) $\sum_\alpha n_a^2 = 18$.

Other Developments

§18 ISOTOPE RULE [7]

In the above example the no. of unknown constants in the potential energy is much greater than the no. of observable frequencies, as is usually the case. The situation is, however, not so hopeless, because the frequencies of the molecule with some of the atoms replaced by their isotopes serve as additional data. The new molecule has evidently the same equilibrium configuration and potential energy as the original one. But in calculating its kinetic energy it must be remembered that all the results of §§1 to 7 are correct, (because they do not concern the masses of the nuclei) but that since now $\begin{pmatrix} \frac{1}{m_1} & & \\ & \ddots & \\ & & \frac{1}{m_N} \end{pmatrix}$ does not necessarily commute with P_c, (21) does not hold. The secular equ. is therefore not factored as before. The vectorial method is still convenient.

If the new kinetic energy matrix is denoted by $\mathscr{G}'$, from (34)

$$\Pi \lambda_i = |\mathfrak{V}\mathscr{G}| = |\mathfrak{V}||\mathscr{G}|,$$
$$\Pi \lambda_i' = |\mathfrak{V}\mathscr{G}'| = |\mathfrak{V}||\mathscr{G}'|.$$

Hence

$$\frac{\Pi \lambda_i}{\Pi \lambda_i'} = \frac{|\mathscr{G}|}{|\mathscr{G}'|} = \frac{\left|\mathscr{L}\begin{pmatrix}\frac{1}{m_1} & & \\ & \ddots & \\ & & \frac{1}{m_N}\end{pmatrix}\mathscr{L}'\right|}{\left|\mathscr{L}\begin{pmatrix}\frac{1}{m_1}' & & \\ & \ddots & \\ & & \frac{1}{m_N}'\end{pmatrix}\mathscr{L}'\right|}.$$

§19 SPLITTING OF HIGH FREQUENCIES [7]

Sometimes it is known that certain force constants are considerably larger than the others, so that the corresponding frequencies of vibration are much greater than the rest. To solve for the low frequencies we may make the approximation of putting the large force constants equal to infinity.

Let

$$\mathscr{G} = \begin{pmatrix} \mathscr{G}_{11} & \mathscr{G}_{12} \\ \mathscr{G}_{21} & \mathscr{G}_{22} \end{pmatrix}, \qquad \mathfrak{V} = \begin{pmatrix} \mathfrak{V}_{11} & \mathfrak{V}_{12} \\ \mathfrak{V}_{21} & \mathfrak{V}_{22} \end{pmatrix}.$$

We want to find the λ's when $\mathfrak{V}_{11} \to \infty$. Now

$$\left|\lambda \mathscr{G}^{-1} \mathfrak{V}^{-1} - I\right| = 0,$$

and when $\mathfrak{V}_{11} \to \infty$, $\mathfrak{V}^{-1} \to \begin{pmatrix} 0 & 0 \\ 0 & \mathfrak{V}_{22}^{-1} \end{pmatrix}$. Since

$$\mathscr{G}^{-1} = \begin{pmatrix} (\mathscr{G}^{-1})_{11} & (\mathscr{G}^{-1})_{12} \\ (\mathscr{G}^{-1})_{21} & (\mathscr{G}_{22} - \mathscr{G}_{21}\mathscr{G}_{11}^{-1}\mathscr{G}_{12})^{-1} \end{pmatrix},$$

we get

$$\left|\lambda \begin{pmatrix} 0 & (\mathscr{G}^{-1})_{12}\mathfrak{V}_{22}^{-1} \\ 0 & (\mathscr{G}_{22} - \mathscr{G}_{21}\mathscr{G}_{11}^{-1}\mathscr{G}_{12})^{-1}\mathfrak{V}_{22}^{-1} \end{pmatrix} - I\right| = 0,$$

or

$$\left|\lambda(\mathscr{G}_{22} - \mathscr{G}_{21}\mathscr{G}_{11}^{-1}\mathscr{G}_{12})^{-1}\mathfrak{V}_{22}^{-1} - I\right| = 0,$$

i.e.

$$\left|\lambda(\mathscr{G}_{22} - \mathscr{G}_{21}\mathscr{G}_{11}^{-1}\mathscr{G}_{12})^{-1} - \mathfrak{V}_{22}\right| = 0.$$

Acknowledgement

The author wishes to express his thanks to Prof. T. Y. Wu for his continual guidance.

++++++++++++++

References

1. von H. Bethe, Ann. d. Physik, 3 (1929), 133.

2. E. Wigner, Göttingen Nachrichton, (1930), p.133.

3. See 1. and 2., also Speiser: Theorie der Gruppen von Endlicher Ordnung (Berlin, 1927).
   and J. E. Rosenthal and G. M. Murphy, Rev. Mod. Phys., 8 (1936), 317.
   The degeneracy and other properties are tabulated by E. B. Wilson Jr. in, J. Chem. Phys., 2 (1934), 432. (complete to molecules containing 7 atoms)
   The irreducible representations are given in F. Seitz, Zeits. f. Krist., A88 (1934), 433.

4. E. Wigner, Gruppentheorie. (Braunschweig 1931) This book is always meant when Wigner is referred to.

5. Wigner, p.86.

6. ibid. p.83.

7. E. B. Wilson Jr., J. Chem. Phys., 9 (1941), 76.

8. J. E. Rosenthal and G. M. Murphy, Rev. Mod. Phys., 8 (1936), 317.

9. E. B. Wilson Jr. and B. L. Crawford Jr., J. Chem. Phys., 8 (1936), 223.

10. D. M. Dennison, Rev. Mod. Phys., 3 (1931), 280.

11. J. B. Howard and E. B. Wilson, Jr., J. Chem. Phys., 2 (1934), 630.

GROUP THEORY AND THE VIBRATION OF POLYATOMIC MOLECULES

Cheng-Ning Yang (楊振寧)

INTRODUCTION Informations about the structure of molecules can always be drawn from the analysis of their vibrational spectra, but owing to the mathematical difficulties involved in the theoretical calculation, only very simple types of molecules can be studied. The method developed by Bethe[1] in 1929, and then more completely by Wigner[2], however, removed considerably this difficulty. It is our purpose here to present the method together with some of the developments after them. A new method of finding the symmetrical coordinates is given (§4), in which the symmetry is preserved from step to step in spite of the existence of redundant coordinates. The theorem in §8 which renders the calculation of the degree of degeneracy very simple is also believed to be new.

The Symmetry of a Molecule

§1 MATHEMATICAL EXPRESSION OF SYMMETRY There are reasons to suppose that the nuclei in a molecule arrange themselves in symmetrical positions when in equilibrium; i.e. some operations (consisting of reflections and rotations) bring the molecule into itself. (For molecules containing isotopes this statement must be slightly modified. cf. §18) If we choose a set of rectangular coordinate axes with the origin at the centre of mass of the molecule in equilibrium, each covering operation C can be represented by an orthogonal matrix Γ_c (order: 3×3) so that the point $\begin{pmatrix}x\\y\\z\end{pmatrix}$ is brought to $\Gamma_c\begin{pmatrix}x\\y\\z\end{pmatrix}$ by the operation. Let $n_1, n_2, \cdots n_n$ be a set of coordinates specifying the relative positions of the nuclei (e.g. the distances between the nuclei and the angles between the bonds) in the molecule. When the nuclei vibrate about their positions of equilibrium, these n's vary (cf. §6). Let $R_1, R_2, \cdots R_n$ be their increments. Further, let $x_1, y_1, z_1, x_2, \cdots x_N, y_N, z_N$ be the increments of the rectangular coordinates of the N nuclei. For small vibrations the R's are linear in the x's, y's and z's:

$$R = \begin{pmatrix}R_1\\R_2\\\vdots\\R_n\end{pmatrix} = B\mathfrak{e}, \quad \text{where } \mathfrak{e} = \begin{pmatrix}x_i\\y_i\\z_i\end{pmatrix}, \qquad (1)$$

B being a constant matrix of n rows and $3N$ columns. Now after

-2-

the operation C, the molecule is indistinguishable from its original self, and we have a new equation obtained by writing (1) down for the new molecule:
$$\overset{c}{R} = B'\overset{c}{e} \qquad (2)$$
Here $\overset{c}{R}$'s are the coordinates of the molecule which will be brought into coincidence with R's by the operation C, and
$$\overset{c}{e} = \begin{pmatrix} T_c \begin{pmatrix} x_{c^{-1}1} \\ y_{c^{-1}1} \\ z_{c^{-1}1} \end{pmatrix} \\ \vdots \\ T_c \begin{pmatrix} x_{c^{-1}N} \\ y_{c^{-1}N} \\ z_{c^{-1}N} \end{pmatrix} \end{pmatrix} \qquad (3)$$
where $c^{-1}i$ is the nucleus which will become after the operation C the nucleus i. Let Z_c be a square matrix of order $3N$ with the elements $_{ix}(Z_c)_{jx} = \delta_{j, c^{-1}i}$, $_{ix}(Z_c)_{jy} = 0$ etc. $i,j = 1,2,\cdots N$, and let T_c stand for $\begin{pmatrix} T_c & & \\ & \ddots & \\ & & T_c \end{pmatrix}$, then (2) and (3) give $\overset{c}{R} = B T_c Z_c e$. (4)

This equation holds for every operation C and is the mathematical expression of the symmetry of the molecule.

EXAMPLE Consider three equivalent nuclei forming an equilateral triangle. Let C be the operation: Rotation counterclockwise through $120°$ about O. Then
$$c^{-1}1 = 3, \quad c^{-1}2 = 1, \quad c^{-1}3 = 2,$$

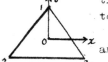

and $Z_c = \begin{pmatrix} 0 & 0 & I \\ I & 0 & 0 \\ 0 & I & 0 \end{pmatrix}$, $T_c = \begin{pmatrix} -\frac{1}{2} & -\frac{\sqrt{3}}{2} & 0 \\ \frac{\sqrt{3}}{2} & -\frac{1}{2} & 0 \\ 0 & 0 & 1 \end{pmatrix}$.

Let R_1 be the increment of the distance $\overline{12}$. $R_1 = \frac{1}{2}x_1 + \frac{\sqrt{3}}{2}y_1 - \frac{1}{2}x_2 - \frac{\sqrt{3}}{2}y_2$.
Then $\overset{c}{R}_1$ is that of $\overline{31}$, and $\overset{c}{R}_2 = -\frac{1}{2}x_2 + \frac{\sqrt{3}}{2}y_1 + \frac{1}{2}x_3 - \frac{\sqrt{3}}{2}y_3$.
Thus $B = (\frac{1}{2}, \frac{\sqrt{3}}{2}, 0, -\frac{1}{2}, -\frac{\sqrt{3}}{2}, 0, 0, 0, 0)$.

(4) becomes the identity
$$-\frac{1}{2}x_2 + \frac{\sqrt{3}}{2}y_1 + \frac{1}{2}x_3 - \frac{\sqrt{3}}{2}y_3 = (\frac{1}{2} \ \frac{\sqrt{3}}{2} \ 0 \ -\frac{1}{2} \ -\frac{\sqrt{3}}{2} \ 0 \ 0 \ 0 \ 0) \begin{pmatrix} T_c & 0 & 0 \\ 0 & T_c & 0 \\ 0 & 0 & T_c \end{pmatrix} Z_c \begin{pmatrix} x_1 \\ y_1 \\ \vdots \end{pmatrix}.$$

§2 FUNDAMENTAL RELATIONSHIP In some instances the coordinates $R_1, R_2, \cdots R_n$ are sufficient to determine $\overset{c}{R}_1, \overset{c}{R}_2, \cdots \overset{c}{R}_n$ for all covering operations C. This is the case if (i) the R's contain only complete sets of equivalent coordinates (e.g. in the last example when $\begin{pmatrix} R_1 \\ R_2 \\ R_3 \end{pmatrix}$ = increment of $\begin{pmatrix} \overline{12} \\ \overline{23} \\ \overline{31} \end{pmatrix}$); or if (ii) the R's are all that are necessary to describe the internal structure of the molecule. In both cases we have for small vibrations $\overset{c}{R} = A_c R$, where A_c is in case (i) an orthogonal matrix having as elements

—3—

0 or 1, and in case (ii) a matrix of order $n \times n$. By (4),
$$B P_c Z_c \mathscr{C} = \hat{R} = A_c R = A_c B \mathscr{C}.$$
But $\mathscr{C}$ is arbitrary (cf. §6), hence
$$B P_c Z_c = A_c B. \qquad (5)$$
This is the fundamental relationship on which all the following deductions are based.

§3 GROUP PROPERTIES To make further developments we notice that the covering operations C form a group and that the R's, Z's and A_c's each form a group isomorphic* with it. The group is known as the "point group". They are of such importance that their properties have been investigated in detail.[3]

Choice of Internal Coordinates

§4 INDEPENDENT REDUCED COORDINATES We first choose the coordinates $R_1, R_2, \cdots R_n$ so that they contain only complete sets of equivalent internal coordinates, and such that they are more than necessary for the determination of the structure of the molecule. The simplest way is to choose the increments of the internuclear distances and the bond angles as the R's. In the example of §1 we may take the increments of the bonds $\overline{12}$, $\overline{23}$ and $\overline{31}$ as R_1, R_2 and R_3; or those of the lengths $\overline{O1}$, $\overline{O2}$, $\overline{O3}$ and the angles $1\hat{O}2$, $2\hat{O}3$ and $3\hat{O}1$ as the R's. The matrix B can now be determined (§§11,12). Evidently our choice belongs to the case (i) of §2, so that the A_c's are orthogonal and have as elements 0 or 1. It is plain that $_i(A_c)_j = 0$ if R_i and R_j are not equivalent. We shall make use of the following theorem in group theory[5]:

If A_c form a group of orthogonal matrices, and W_c^{α} ($\alpha = 1, 2, \cdots k$) are the irreducible orthogonal representations of the group, there exists an orthogonal matrix M such that $W_c = M A_c M'$ is of the form
$$\begin{pmatrix} W_c^1 & & & & 0 \\ & W_c^1 & & & \\ & & W_c^2 & & \\ & & & \ddots & \\ 0 & & & & W_c^k \end{pmatrix} \qquad (6)$$

* Let C'C be the resultant operation of first operating C and then C', we have
$$P_{c'c} = P_{c'} P_c, \quad Z_{c'c} = Z_{c'} Z_c \quad \text{but} \quad A_{c'c} = A_c A_{c'}.$$

-4-

We define* $Q = \begin{pmatrix} Q_1 \\ Q_2 \\ \vdots \\ Q_n \end{pmatrix} = MR$ as the "reduced coordinates". Evidently
$$\tilde{Q} = M\tilde{R} = MA_c R = MA_c M^{-1} Q = W_c Q.$$
Now not all the Q's are independent. To select out the independent ones we need the following theorem:

THEOREM It is always possible to drop out some of the Q's so that (i) the remaining ones are all independent,

(ii) the dropped ones depend on the remaining ones,

and (iii) the remaining ones belong to complete blocks of the group of matrices W_c.

Because of the properties (i) and (ii), the remaining coordinates $\mathcal{R} = \begin{pmatrix} \mathcal{R}_1 \\ \mathcal{R}_{3N-6} \end{pmatrix}$ are called the "independent reduced coordinates", and from (iii), $\tilde{\mathcal{R}} = \alpha_c \mathcal{R}$ where α_c is obtained from W_c by dropping some blocks. These coordinates may also be obtained in other ways, and have been given various names by different authors. They are the "internal symmetry coordinates" of Wilson[7] and the "geometrical symmetry coordinates" of Rosenthal and Murphy[8]. The relation between $\mathcal{R}$ and the "symmetry coordinates" of Howard and Wilson[9] will be given in §14.

PROOF OF THE THEOREM Since Q is not entirely arbitrary, there exists a matrix H of order $\nu \times n$ so that $HH' = I$, $HQ = 0$, (7)
where I is the unit matrix and the ' stands for "transposed". We may prove that the rank of H is ν and that there exists a matrix H_1 of order $(n-\nu) \times n$ such that $\binom{H}{H_1}$ is orthogonal. Put
$$\binom{H}{H_1} Q = \binom{0}{Q_1}, \text{ then } Q = (H' H_1') \binom{0}{Q_1} = H_1' Q_1. \quad (8)$$
We may suppose that $\nu = n-(3N-6)$ so that Q_1 is arbitrary. The equ. $HQ = 0$ must be invariant under an operation C:
$$0 = H\tilde{Q} = HW_c Q = HW_c H_1' Q_1.$$
Hence $HW_c H_1' = 0.$

Thus $\binom{H}{H_1} W_c (H' H_1') = \begin{pmatrix} HW_c H' & 0 \\ H_1 W_c H' & H_1 W_c H_1' \end{pmatrix}.$

This matrix is orthogonal (for $\binom{H}{H_1}$ and W_c are both orthogonal). Hence
$$\binom{H}{H_1} W_c (H' H_1') = \begin{pmatrix} HW_c H' & 0 \\ 0 & H_1 W_c H_1' \end{pmatrix}. \quad (9)$$
Thus $HW_c H'$ is orthogonal and is a group isomorphic with the one formed by C; i.e. $(HW_c H')(HW_{c'} H') = HW_{cc'} H'.$

*We assume here that Q is real. There is probably no difficulty in handling complex coordinates, but they are not necessary unless molecules of symmetry C_n, C_{nh} or S_n ($n>2$) are studied.

-5-

By the theorem quoted above there exists an orthogonal matrix J such that $JW_cH'J' = \omega_c$ is of the form (6); i.e. if the "untermatrices" of W_c are $_{\alpha\beta}(W_c)_{\alpha'\beta'} = \delta_{\alpha\alpha'}\delta_{\beta\beta'}W_c^\alpha$, $\alpha = 1,2,\ldots k$, $\beta = 1,2,\ldots n_\alpha$, those of ω_c must be $_{ab}(\omega_c)_{a'b'} = \delta_{aa'}\delta_{bb'}W_c^a$, $a = 1,2,\ldots k$, $b = 1,2,\ldots \mu_a$. By (9), the irreducible blocks of ω_c must all be that of W_c, hence $\mu_a \leq n_a$. Now $(JH)W_c = JHW_c\binom{H'H_1'}{H_1} = (JW_cH' \; 0)\binom{H}{H_1} = \omega_cJH$ (10)

i.e. $_{ab}(JH)_{\alpha\beta}W_c^\alpha = W_c^a(JH)_{\alpha\beta}$.

But W_c^α is irreducible, hence $_{ab}(JH)_{\alpha\beta} = \delta_{a\alpha}\,_b\lambda_\beta^\alpha I$, (11)

where $_b\lambda_\beta^\alpha$ is an ordinary number. Since JH is of rank ν, the submatrix $(_b\lambda_\beta^\alpha)$, $b = 1,2,3,\ldots \mu_\alpha$, $\beta = 1,2,3,\ldots n_\alpha$ is of rank μ_α. Thus for every α there exists a set K_α of μ_α integers all $\leq n_\alpha$, such that the square matrix $(_b\lambda_\beta^\alpha)$, $b = 1,2,\ldots \mu_\alpha$, β in K_α is nonsingular. Hence JH may be divided into two untermatrices, one (of order $\nu \times \nu$) composed of those untermatrices (11) for which β is in the set K_α, and the other those for which β is not in K_α. The former is evidently nonsingular. Let H_2 and H_3 be the corresponding untermatrices of $H = J^{-1}(JH)$. Evidently a transposition of columns bring may H into the form $(H_2 \; H_3)$. Now the constraint on Q is $HQ = 0$. Hence a corresponding transposition of the rows of Q bring it to $\binom{Q_2}{Q_3}$ so that

$$H_2Q_2 + H_3Q_3 = 0.$$

Since H_2 is nonsingular, Q_3 may be chosen as the independent variable, Q_2 being dependent on it. Thus the latter can be dropped and the conditions (i), (2) and (iii) are satisfied.

§5 CALCULATION OF $\text{Spur}(\alpha_c)$. Suppose that the constraints on the R's are given by $\mathcal{D}R = 0$ where $\mathcal{D}\mathcal{D}' = I$. (12)

THEOREM $\text{Spur}(\alpha_c) = \text{Spur}(A_c)' - \text{Spur}(\mathcal{D}A_c\mathcal{D}')$ (13)

This theorem makes the calculation of $\text{Spur}(\alpha_c)$ very simple (because the elements of A_c are 0 or 1), and before the transformation from R to Q is carried out. Also it enables us to calculate the contributions of the different constraints separately.

PROOF The constraints on $R_1, R_2, \ldots R_n$ are $\mathcal{D}R = 0$. Hence we may take the matrix H of (7) to be $\mathcal{D}M'$. With the notation used there we have $\hat{Q}_1 = H_1\hat{Q} = H_1W_cQ = H_1W_cH_1'Q_1$. (14)

Since the R's are all independent, there exists M_1 such that $Q_1 = M_1R$. Hence $\hat{Q}_1 = M_1\hat{R} = M_1\alpha_cR = M_1\alpha_cM_1^{-1}Q_1$.

On comparison with (14) it follows that $H_1W_cH_1' = M_1\alpha_cM_1^{-1}$. Thus

$$\text{Spur}(\alpha_c) = \text{Spur}(H_1W_cH_1') = \text{Spur}\left[\binom{H}{H_1}W_c(H' \; H_1')\right] - \text{Spur}(HW_cH') = \text{Spur}(A_c) - \text{Spur}(\mathcal{D}A_c\mathcal{D}').$$

EXAMPLE Consider the molecule CH_3Cl.

-6-

EXAMPLE Consider the molecule CH_3Cl. Take

R_1, R_2, R_3, R_4 = increments of the distances $\overline{CCl}, \overline{CH_\alpha}, \overline{CH_\beta}, \overline{CH_\gamma}$,

R_5, R_6, R_7 = increments of the angles $Cl\hat{C}H_\alpha, Cl\hat{C}H_\beta, Cl\hat{C}H_\gamma$,

R_8, R_9, R_{10} = increments of the angles $H_\beta\hat{C}H_\gamma, H_\gamma\hat{C}H_\alpha, H_\alpha\hat{C}H_\beta$.

The constraint is $[R_5+R_6+R_7]f + [R_8+R_9+R_{10}] = 0$,
where f is a constant. Thus $\mathcal{D} = \frac{1}{\sqrt{3f^2+3}}(0\ 0\ 0\ 0\ f\ f\ f\ 1\ 1\ 1)$.
The theorem leads to

Operation C		Spur(A_c)	Spur($\mathcal{D}A_c\mathcal{D}'$)	Spur(α_c)
No Motion : C_1		10	1	9
Rotation about $\overline{CCl}$ through	120°: C_2	1	1	0
	240°: C_3			
Reflection about the plane	$ClCH_\alpha : C_4$	4	1	3
	$ClCH_\beta : C_5$			
	$ClCH_\gamma : C_6$			

The Kinetic and the Potential Energies

§6 CHOICE OF AXES Suppose that $\begin{pmatrix} R_1 \\ R_2 \\ \vdots \\ R_{3N-6} \end{pmatrix} = R$ are the increments of $3N-6$ independent internal coordinates. For small vibrations, $R = B\mathcal{C}$ where $\mathcal{C} = \begin{pmatrix} x_1 \\ \vdots \\ x_{3N} \end{pmatrix}$ is defined in §1. Suppose that the equilibrium positions of the nuclei are, in rectangular coordinates, $(\bar{x}_1, \bar{Y}_1, \bar{Z}_1), (\bar{x}_2, \bar{Y}_2, \bar{Z}_2), \cdots (\bar{x}_N, \bar{Y}_N, \bar{Z}_N)$. Let $m_1, m_2, \ldots m_N$ be their masses. Write

$$F = \begin{pmatrix} m_1 & 0 & 0 & m_2 & 0 & 0 & - & - & - & 0 \\ 0 & m_1 & 0 & 0 & m_2 & 0 & & & & 0 \\ 0 & 0 & m_1 & 0 & 0 & m_2 & & & & m_N \\ 0 & m_1\bar{z}_1 & -m_1\bar{Y}_1 & 0 & m_2\bar{z}_2 & -m_2\bar{Y}_2 & & & & m_N\bar{Y}_N \\ -m_1\bar{z}_1 & 0 & m_1\bar{X}_1 & -m_2\bar{z}_2 & 0 & m_2\bar{X}_2 & & & & m_N\bar{X}_N \\ m_1\bar{Y}_1 & -m_1\bar{X}_1 & 0 & m_2\bar{Y}_2 & -m_2\bar{X}_2 & 0 & & & & 0 \end{pmatrix}$$

Then the first column of $B\begin{pmatrix} \frac{1}{m_1} & & \\ & \ddots & \\ & & \frac{1}{m_N} \end{pmatrix}F'$ is the row matrix $B\begin{pmatrix} 1 \\ 0 \\ 0 \\ \vdots \\ 0 \end{pmatrix}$ which is the value of R when all the nuclei are displaced by 1 unit of length along the X-axis. But the R's are the increments of internal variables, hence $B\begin{pmatrix} 1 \\ 0 \\ \vdots \\ 0 \end{pmatrix} = 0$. In the same way we can show that

$$B\begin{pmatrix} \frac{1}{m_1} & & \\ & \ddots & \\ & & \frac{1}{m_N} \end{pmatrix}F' = 0.$$

We have thus far described the molecule in a certain system of rectangular coordinates. But when the molecule moves and rotates in space as well as vibrates, there remains an arbitrariness

* From this it is immediately seen that any constraint having the same coefficient for equivalent coordinates contributes 1 to Spur($\mathcal{D}A_c\mathcal{D}'$).

in fixing the coordinate axes to the molecule. The following method of choosing these axes is, however, the most preferable.
The $3N-6$ internal variables $R_1, R_2, \ldots R_{3N-6}$ is determined uniquely from the structure of the molecule. We define $x_1, y_1, \ldots z_N$ by

$$\mathscr{C} = \begin{pmatrix} x_1 \\ \vdots \\ z_N \end{pmatrix} = \begin{pmatrix} B \\ F \end{pmatrix}^{-1} \begin{pmatrix} R \\ 0 \end{pmatrix}, \tag{15}$$

so that
$$F\mathscr{C} = 0. \tag{16}$$

From the definition of B, we conclude that a system of axes of reference can be found so that the positions of the nuclei are $(X_1+x_1, Y_1+y_1, Z_1+z_1), (X_2+x_2, Y_2+y_2, Z_2+z_2), \ldots (X_N+x_N, Y_N+y_N, Z_N+z_N)$ when the molecule is not very much distorted from its equilibrium structure. We have thus 6 external variables specifying the position and orientation of the axes in space and $3N-6$ variables $R_1, R_2, \ldots R_{3N-6}$ specifying by means of (15) the positions of the nuclei with respect to these axes.

§7 THE KINETIC ENERGY IN TERMS OF R

Equ.(16) expresses the facts that the origin of our moving axes is at any time the centre of mass of the molecule, and that the moment of momentum of the molecule in this system of reference is of the order of $mx\dot{x}$. Thus if the molecule rotates in space with an angular velocity ω about its centre of mass, its kinetic energy is

$$T_{\text{centre of mass}} + T_{\text{rotation}} + \tfrac{1}{2} \sum m_i (\dot{x}_i^2 + \dot{y}_i^2 + \dot{z}_i^2) + \text{term} \sim mx\dot{x}\omega.$$

But $T_{\text{rotation}} = (\text{K.E. of rotation if } \mathscr{C}=0) + \text{term} \sim m\omega^2 xX$.

Now $T_{\text{centre of mass}} + (\text{K.E. of rotation if } \mathscr{C}=0)$ depends on the external variables only (together, of course, with their time derivatives), and in a gas, due to thermal agitation, $\omega X \sim \dot{x}$.

Thus in the first approximation the equ. of motion is to be derived from (cf. §§15,16)
$$T = \tfrac{1}{2} \sum m_i (\dot{x}_i^2 + \dot{y}_i^2 + \dot{z}_i^2) = \tfrac{1}{2} (\dot{R}' 0) \begin{pmatrix} B \\ F \end{pmatrix}^{-1\prime} \begin{pmatrix} m_1 \\ & \ddots \\ & & m_N \end{pmatrix} \begin{pmatrix} B \\ F \end{pmatrix}^{-1} \begin{pmatrix} \dot{R} \\ 0 \end{pmatrix}.$$

But
$$\begin{pmatrix} B \\ F \end{pmatrix} \begin{pmatrix} \tfrac{1}{m_1} \\ & \ddots \\ & & \tfrac{1}{m_N} \end{pmatrix} (B' F') = \begin{pmatrix} B\left(\tfrac{1}{m_i} \ddots \tfrac{1}{m_N}\right)B' & 0 \\ 0 & F\left(\tfrac{1}{m_i} \ddots \tfrac{1}{m_N}\right)F' \end{pmatrix}.$$

Hence writing
$$G = B\left(\tfrac{1}{m_i} \ddots \tfrac{1}{m_N}\right)B', \tag{17}$$

we have
$$2T = \dot{R}' G^{-1} \dot{R}. \tag{18}$$

§8 THE KINETIC ENERGY IN TERMS OF $\mathcal{R}$

If we take the independent reduced coordinates $\mathcal{R}$ (§4) to be the R's of the last section, the results may be summarized:

$$\mathcal{R} = \mathscr{L}\mathscr{C}, \qquad \mathcal{G} = \mathscr{L}\left(\tfrac{1}{m_i} \ddots \tfrac{1}{m_N}\right)\mathscr{L}', \qquad 2T = \dot{\mathcal{R}}' \mathcal{G}^{-1} \dot{\mathcal{R}}. \tag{19}$$

Now by (5), $\mathscr{L} P_c z_c = \alpha_c \mathscr{L}$ (α stands here for A_c, cf. §4)

-8-

So that $\quad \mathcal{Q} = \mathcal{L}\left(\frac{1}{m_1}\cdots\frac{1}{m_N}\right)\mathcal{L}' = \alpha_c'P_c Z_c\left(\frac{1}{m_1}\cdots\frac{1}{m_N}\right)Z_c'P_c'\mathcal{L}'\alpha_c$.

But $\left(\frac{1}{m_1}\cdots\frac{1}{m_N}\right)$, P_c and Z_c commute* with each other, hence

$$\mathcal{Q} = \alpha_c'\mathcal{L}\left(\frac{1}{m_1}\cdots\frac{1}{m_N}\right)\mathcal{L}'\alpha_c = \alpha_c'\mathcal{Q}\alpha_c.$$

Suppose $\quad {}_{\alpha\beta}(\alpha_c)_{\alpha'\beta'} = \delta_{\alpha\alpha'}\delta_{\beta\beta'}W_c^\alpha,\quad \alpha=1,2,\cdots k,\ \beta=1,2,\cdots n_\alpha,\quad (20)$

where W_c^α is irreducible. (Ch. 4) Since $\alpha_c\mathcal{Q} = \mathcal{Q}\alpha_c$, we have

$$W_c^\alpha {}_{\alpha\beta}(\mathcal{Q})_{\alpha'\beta'} = {}_{\alpha\beta}(\mathcal{Q})_{\alpha'\beta'} W_c^{\alpha'}.$$

Hence[6] $\quad {}_{\alpha\beta}(\mathcal{Q})_{\alpha'\beta'} = \delta_{\alpha\alpha'}\delta_{\beta_p\beta_{p'}}g_p^\alpha I.\quad (21)$

If W_c^α is of order d_α, this shows that the nonvanishing elements of $\mathcal{Q}$ are in $\sum_{\alpha=1}^{k} d_\alpha$ diagonal blocks of which d_α are identical of dimension $n_\alpha \times n_\alpha$.

§9 THE POTENTIAL ENERGY IN TERMS OF $\mathcal{R}$ The potential energy depends on the internal coordinates only. For small vibrations, it is approximately equal to $\frac{1}{2}\mathcal{R}'\mathcal{U}\mathcal{R}$, where $\mathcal{U}$ is a positive symmetrical matrix, because when in equilibrium the molecule has a minimum potential energy. The covering operations leave the potential energy unchanged: $\frac{1}{2}\mathcal{R}'\mathcal{U}\mathcal{R} = \frac{1}{2}\tilde{\mathcal{R}}'\mathcal{U}\tilde{\mathcal{R}} = \frac{1}{2}\mathcal{R}'\alpha_c'\mathcal{U}\alpha_c\mathcal{R}$.

But $\mathcal{R}$ is arbitrary, hence $\quad \mathcal{U} = \alpha_c'\mathcal{U}\alpha_c$

Thus we have, just as we had (21),

$$ {}_{\alpha\beta}(\mathcal{U})_{\alpha'\beta'} = \delta_{\alpha\alpha'}\delta_{\beta\beta'}v_p^\alpha I,\quad \alpha=1,2,\cdots k,\ \beta=1,2,\cdots n_\alpha.\quad (22)$$

There are therefore totally $\sum_{\alpha=1}^{k} n_\alpha^2$ independent constants in the potential energy, which are usually unknowns."

Calculation of the Kinetic Energy

§10 THE TRANSFORMATION MATRIX M[7]   In §4 the orthogonal matrix M was introduced to reduce A_c. Since

$$A_c = \begin{pmatrix} A_{c1} & 0 \\ 0 & A_{c2} \ddots \end{pmatrix}$$

where each A_{cp} belongs to a set of equivalent coordinates, (§4,P.3)

$$M = \begin{pmatrix} M_1 & 0 \\ 0 & M_2 \ddots \end{pmatrix}$$

where M_p is orthogonal. Hence $\widetilde{W}_{cp} = M_p A_{cp} M_p'$ is of the form (6), i.e. $\quad \widetilde{W}_{cp} = \begin{pmatrix} W_{cp}^1 & & \\ & W_{cp}^1 & \\ & & \ddots W_{cp}^k \end{pmatrix}$.

Suppose that W_{cp}^1 is of dimension $d \times d$. Denote the first d rows of M_p by $\quad L = \begin{pmatrix} {}_1(M_p)_1, & {}_1(M_p)_2, & \cdots \\ \vdots & & \\ {}_d(M_p)_1, & \cdots & {}_d(M_p)_{g_p} \end{pmatrix}$,

where $g_p = $ no. of equivalent coordinates $\mathcal{R}$ of the p-th set.

Now $\quad \widetilde{W}_{cp} M_p = M_p A_{cp}$

Hence $\quad W_{cp}^1 L = L A_{cp}$

Thus $\quad L'L = A_{cp}'L'L A_{cp}$ i.e. $(L'L)A_{cp} = A_{cp}(L'L)$.

* Cf. §18

Hence
$$\sum_{\tau=1}^{g_p}{}_\tau(L'L)_\tau {}_\tau(A_{cp})_2 = \sum_{\tau=1}^{g_p}{}_1(A_{cp})_\tau {}_\tau(L'L)_2$$

There exists an operation C which brings the second coordinate of the p-th set to the first, for which $\tilde{R}_1 = R_2$ so that ${}_1(A_{cp})_\tau = \delta_{\tau 2}$, ${}_\tau(A_{cp})_2 = \delta_{\tau 1}$. Hence ${}_1(L'L)_1 = {}_2(L'L)_2$. Thus all diagonal elements of $L'L$ are equal, and $\sum_{\tau=1}^{3}{}_\tau(M_p)^2_\tau = {}_\tau(L'L)_\tau = \frac{1}{g_p}\text{Spur}(L'L) = \frac{1}{g_p}\text{Spur}(LL') = \frac{d}{g_p}$.

Now d can be determined from the values of $\text{Spur}(\alpha_k)$, (cf. equ. (30)) so that this relation facilitates the calculation of M.

§11 VECTORIAL NOTATION[7] The matrix $\mathcal{J} = \mathcal{L}\left(\frac{1}{m_1}\cdots\frac{1}{m_N}\right)\mathcal{L}'$ is calculated by first computing $\mathcal{L}$. Now $\mathcal{L}$ is a submatrix of MBQ. Hence

$$\mathcal{L} = M_1 B \qquad \text{where} \quad M_1 = \text{some rows}(M)_{\text{all columns}}. \quad (23)$$

Thus we have to find B first.

Now $\qquad R = B \cdot \mathcal{E}$.

Write $(x_\tau, y_\tau, z_\tau) = \vec{\mathcal{E}}_\tau$, $({}_\kappa B_{\tau x}, {}_\kappa B_{\tau y}, {}_\kappa B_{\tau z}) = \vec{{}_\kappa B}_\tau$.

Then $R_\kappa = \sum_\tau \vec{{}_\kappa B}_\tau \cdot \vec{\mathcal{E}}_\tau$. In a similar way we shall write

$${}_{\alpha\rho\gamma}(\mathcal{L})_\tau = \sum_{\kappa=1}^{N}{}_{\alpha\rho\gamma}(M_1)_\kappa \vec{{}_\kappa S}_\tau = \vec{{}_{\alpha\rho\gamma}S}_\tau. \quad \alpha=1,2,\cdots d_\alpha, \tau=1,2\cdots N. \text{ (cf.(21))} \quad (24)$$

§12 EXPLICIT EXPRESSION OF $\vec{{}_\kappa S}_\tau$[?] The use of the vectors $\vec{S}$ is advantageous because they are independent of the coordinate axes. Two kinds of R's are commonly used:

(i) $R =$ the increment of the bond between the nuclei t' and t'' in length. Evidently when $\vec{\mathcal{E}}_{t'} = \vec{\mathcal{E}}_{t''} = 0$, $R = 0$. Hence $\vec{{}_t S}_\tau = 0$ for all t except t' or t''.

If $0 = y_{t'} = z_{t'} = x_{t''} = y_{t''} = z_{t''}$, we have $R = x_{t'}\cos\theta$, so that $B_{\kappa x} = \cos\theta$.

Hence $\begin{cases} \vec{S}_{t'} = \text{unit vector from } t'' \text{ to } t'. \\ \vec{S}_{t''} = -\vec{S}_{t'}. \\ \text{all other } \vec{S}_t \text{ are } 0. \end{cases}$ (25)

(ii) $R =$ the increment of the angle $t'-t-t''$. Evidently all $\vec{S}$ are 0 except ${}_t\vec{S}_t$, $\vec{S}_{t'}$ and $\vec{S}_{t''}$. Suppose that

$$\vec{S}_{t'} = x'\vec{\zeta} + y'\vec{\eta} + z'\vec{\zeta'}.$$

When $\vec{\mathcal{E}}_t = \vec{\mathcal{E}}_{t''} = 0$, $\vec{\mathcal{E}}_{t'} = \vec{\zeta}$, we have, since t' alone is displaced, $R = \frac{1}{\ell}$. Hence $x' = \frac{1}{\ell}$. Similarly by considering the case when $\vec{\mathcal{E}}_t = \vec{\mathcal{E}}_{t''} = 0$, $\vec{\mathcal{E}}_{t'} = \vec{\eta}$, we get $y' = 0$. Proceeding in this way we arrive finally at the first line of (26). Now R vanishes when $\vec{\mathcal{E}}_{t'} = \vec{\mathcal{E}}_t = \vec{\mathcal{E}}_{t''}$. Thus $(\vec{S}_{t'} + \vec{S}_{t''} + \vec{S}_t) = 0$.

-10-

so that
$$\begin{cases} \vec{S}_{k'} = \frac{1}{\epsilon}\vec{3}', & \vec{S}_{k''} = \frac{1}{\gamma}\vec{3}'' \\ \vec{S}_k = -(\vec{S}_{k'} + \vec{S}_{k''}) \\ \text{all other } \vec{S} = 0 \end{cases} \quad (26)$$

From (25) and (26) we can write down, for all κ and t, the vector $_\kappa\vec{S}_t$ from which $\vec{S}$ may be calculated through (24).

§13 **KINETIC ENERGY IN TERMS OF THE $\vec{S}$'S**[7] We shall show however, that not all the $\vec{S}$'s are needed for the calculation of $\mathcal{G}$.

Take (5) $\qquad \mathscr{L} P_c z_c = \mathscr{L}_c t. \qquad (27)$

By (20) $\qquad \sum_{t''}{}_{\alpha\beta}\mathcal{L}_{t''}^\alpha I_c \delta_{t,c't'} = W_c^\alpha{}_{\alpha\beta}\mathcal{L}_t^\alpha$. ($\tau_c$ is defined in §2)

i.e. $\qquad _{\alpha\beta}\mathcal{L}_{ct}^\alpha I_c = W_c^\alpha{}_{\alpha\beta}\mathcal{L}_t^\alpha$.

Hence for any ρ and ρ', $I_c'{}_{\alpha\beta}\mathcal{L}_{ct}^{\alpha'}{}_{\alpha\beta}\mathcal{L}_{ct}^\alpha I_c = {}_{\alpha\beta}\mathcal{L}_t^{\alpha'}{}_{\alpha\beta}\mathcal{L}_t^\alpha$,

so that $\qquad \text{Spur}({}_{\alpha\beta}\mathcal{L}_{ct}^{\alpha'}{}_{\alpha\beta}\mathcal{L}_{ct}^\alpha) = \text{Spur}({}_{\alpha\beta}\mathcal{L}_t^{\alpha'}{}_{\alpha\beta}\mathcal{L}_t^\alpha). \qquad (28)$

Now by (21), $\qquad {}_\rho g_{\rho'}^\alpha \cdot I = {}_{\alpha\beta}g_{\alpha\beta'} = \sum{}_{\alpha\beta}(\frac{1}{m_p}I)_{\alpha\beta'}{}_p. \qquad (29)$

where each p refer to a set of equivalent nuclei.

From (27) $\qquad _{\alpha\beta}(\mathcal{L}_t^\alpha)_p \, {}_p(z)_{pp}(z_c)_p = W_c^\alpha{}_{\alpha\beta}(\mathcal{L}_t)_p$.

Hence $\qquad _{\alpha\beta}(\mathcal{L}_t^\alpha)_p\,{}_{\alpha\beta}\mathcal{L}_p' = W_c^\alpha{}_{\alpha\beta}\mathcal{L}_p^{\alpha'}$,

i.e. $\qquad ({}_{\alpha\beta}\mathcal{L}_p^\alpha\,{}_{\alpha\beta}\mathcal{L}_p')W_c^\alpha = W_c^\alpha({}_{\alpha\beta}\mathcal{L}_p^\alpha\,{}_{\alpha\beta}\mathcal{L}_p')$.

Thus[6] $\qquad _{\alpha\beta}\mathcal{L}_p^\alpha\,{}_{\alpha\beta}\mathcal{L}_p' = {}_\rho t_{\rho'}^{\alpha\rho} I, \qquad (30)$

where ${}_\rho t_{\rho'}^{\alpha\rho}$ is an ordinary number. ~~Substitute this into (29)~~ Hence

$${}_\rho t_{\rho'}^{\alpha\rho} d_\alpha = \text{Spur}({}_{\alpha\beta}\mathcal{L}_p^\alpha\,{}_{\alpha\beta}\mathcal{L}_{p'}') = \text{Spur}({}_{\alpha\beta}\mathcal{L}_p^{\alpha'}\,{}_{\alpha\beta}\mathcal{L}_p) = \sum_{t \text{ in } p}\text{Spur}({}_{\alpha\beta}\mathcal{L}_t^{\alpha'}\,{}_{\alpha\beta}\mathcal{L}_t^\alpha) \quad (31)$$

From (29) we have,
$${}_\rho g_{\rho'}^\alpha I = \sum_p \frac{1}{m_p}{}_\rho t_{\rho'}^{\alpha\rho} I. \qquad (32)$$

Now consider the sum of (28). Since ct goes over all values of t in p when c goes over all the operations of the group,

$$\text{Spur}({}_{\alpha\beta}\mathcal{L}_{ct}^{\alpha'}\,{}_{\alpha\beta}\mathcal{L}_{ct}^\alpha) = \text{Spur}({}_{\alpha\beta}\mathcal{L}_{ct}^{\alpha'}\,{}_{\alpha\beta}\mathcal{L}_{ct}^\alpha)$$

if t and t' are in the same p-th set. Thus (32) becomes

$${}_\rho t_{\rho'}^{\alpha\rho} d_\alpha = g_p \text{Spur}({}_{\alpha\beta}\mathcal{L}_t^{\alpha'}\,{}_{\alpha\beta}\mathcal{L}_t^\alpha)$$

where t represents any nucleus of the p-th set. Substitute into (31) and make use of (24). We obtain:

$${}_\rho g_{\rho'}^\alpha = \frac{1}{d_\alpha}\sum_p \frac{g_p}{m_p}\sum_{\gamma=1}^{d_\alpha}{}_{\alpha\beta}\vec{S}_t \cdot {}_{\alpha\beta'}\vec{S}_t' \qquad (33)$$

Thus in calculating $\mathcal{G}$ (${}_{\alpha\beta}g_{\alpha\beta'} = \sum_\alpha {}_\rho g_{\rho'}^\alpha I$) we have only to know ${}_{\alpha\beta}\vec{S}_k$ for all $\alpha(=1,2,\cdots k)$, $\beta(=1,2,\cdots n_d)$ and $\gamma(=1,2,\cdots d_\alpha)$; but only for one t in each p. From (24) we see that not all $\vec{S}$ are needed.

Secular Equation and Degeneracy

§14 **NORMAL COORDINATES** In terms of the independent reduced coor-

-11-

dinates $\mathcal{R}$ defined in §4, the kinetic energy and the potential energy are $\frac{1}{2}\dot{\mathcal{R}}'\mathcal{G}^{-1}\dot{\mathcal{R}}$ and $\frac{1}{2}\mathcal{R}'\mathcal{H}\mathcal{R}$, (§§8,9) where $\mathcal{G}$ is positive definite. Let $\lambda_1, \lambda_2, \ldots \lambda_{3N-6}$ be the roots of the equ.

$$|\lambda \mathcal{G}^{-1} - \mathcal{H}| = 0, \quad \text{i.e.} \quad |\lambda I - \mathcal{H}\mathcal{G}| = 0. \tag{34}$$

Then there exists a matrix L such that

$$L\mathcal{G}L' = I. \quad L'^{-1}\mathcal{H}L^{-1} = \begin{pmatrix} \lambda_1 & & 0 \\ & \ddots & \\ 0 & & \lambda_{3N-6} \end{pmatrix} = \Lambda.$$

Put $\mathcal{H} = \begin{pmatrix} \eta_1 \\ \vdots \\ \eta_{3N-6} \end{pmatrix} = L\mathcal{R}$, we get immediately

$$2 \times \text{kinetic energy} = \dot{\mathcal{R}}'\mathcal{G}^{-1}\dot{\mathcal{R}} = \dot{\eta}'\dot{\eta} = \sum_{i=1}^{3N-6} \dot{\eta}_i^2, \tag{35}$$

$$2 \times \text{potential energy} = \mathcal{R}'\mathcal{H}\mathcal{R} = \eta'\Lambda\eta = \sum \lambda_i \eta_i^2. \tag{36}$$

These coordinates η_i are known as the "normal coordinate"'s. They are obtained by first solving (34), thus getting λ_i; and then determining L from

$$L\mathcal{G}\mathcal{H} = \Lambda L. \tag{37}$$

This L must be normalized by $L\mathcal{G}L' = I$. \hfill (38)

The normal coordinates are then calculated from $\eta = L\mathcal{R}$.

Now we shall be able to see the rôle that symmetry plays in simplifying the calculations. Equs. (21) and (22) shows the secular equ.(34) is factored into $\sum_{\alpha=1}^{k} d_\alpha$ equations of which d_α are identical and are of the n_α-th degree. The labor of solving for the λ's is considerably saved. Moreover, the no. of unknown constants in $\mathcal{H}$ is reduced.

The normal coordinates are also a special form of the "symmetry coordinates" introduced by Howard and Wilson.[11] The most general form of the symmetry coordinates $\mathfrak{M}$ is given by $\eta = \mathcal{L}\mathfrak{M}$ where

$$\alpha\beta d^\alpha_{\alpha'\beta'} = \delta_{\alpha\alpha'} \beta\rho^\alpha_{\beta'} U^\alpha, \quad \alpha = 1,2,\ldots k, \quad \beta = 1,2,\ldots n_\alpha,$$

in which U^α is orthogonal and $\rho^\alpha_{\beta\beta'}$ form an orthogonal matrix when β and β' range over $1,2,\ldots n_\alpha$.

§15 SOLUTION OF THE PHYSICAL PROBLEM —— CLASSICAL THEORY

The Lagrangian is, from (35) and (36),

$$\tfrac{1}{2}\sum_{i=1}^{3N-6} (\dot{\eta}_i^2 - \lambda_i \eta_i^2).$$

The equs. of motion are therefore

$$\ddot{\eta}_i + \lambda_i \eta_i = 0, \quad i = 1,2,\ldots 3N-6.$$

Hence $\eta_i = (\eta_i)_0 \cos(\sqrt{\lambda_i}\, t + \phi_i)$.

The frequencies of vibration are thus

$$\frac{\sqrt{\lambda_1}}{2\pi}, \frac{\sqrt{\lambda_2}}{2\pi}, \ldots \frac{\sqrt{\lambda_{3N-6}}}{2\pi}.$$

This is correct only in the first approximation; but it serves to give almost all of our present knowledge about the forces within the molecules.

§16 SOLUTION OF THE PHYSICAL PROBLEM —— QUANTUM THEORY Now

$$T = \frac{1}{2} \sum \dot{\eta}_i^2, \qquad V = \frac{1}{2} \sum \lambda_i \eta_i^2.$$

The wave equation is most easily obtained from the variational formulation of the problem:

$$\delta \int \{\tfrac{\hbar^2}{2} g^{ij} \frac{\partial \psi^*}{\partial \eta_i} \frac{\partial \psi}{\partial \eta_j} + (V-E)\psi^*\psi\} \sqrt{g}\, d\eta_1 \cdots d\eta_{3N-6} = 0,$$

where g^{ij} is given by $T = g_{ij}\dot{x}^i\dot{x}^j$. Hence

$$\sum \tfrac{1}{2}(-\hbar^2 \tfrac{\partial^2}{\partial \eta_i^2} + \lambda_i \eta_i^2)\psi = E\psi.$$

The electric moment in any direction is, in the first approximation

$$D_0 + \sum_{i=1}^{3N-6} D_i \eta_i$$

where D_i is immediately calculable from (15). Thus the wave function and selection rules, hence the frequencies in the vibrational spectra of the molecule, are the same as those of $3N-6$ independent harmonic oscillators with coordinates $\eta_1, \eta_2, \ldots \eta_{3N-6}$. The frequencies are therefore exactly those obtained classically.

§17 THE DEGREE OF DEGENERACY

We have seen that the secular equ. (34) is factorized into d_1 identical equs. of the n_1-th degree, d_2 identical equs. of the n_2-th degree, ... There must therefore be n_i frequencies each corresponding to d_i different normal moods of vibration (d_i-fold degenerate). The nos. $d_1, d_2 \ldots$ are accordingly called "the degrees of degeneracy".

It is interesting that they can be determined together with $n_1, n_2, \ldots$ without carrying out the calculation of the normal coordinates. For, by (20)

$$\sum n_d\, \mathrm{Spur}(W_c^d) = \mathrm{Spur}(\mathcal{O}_c).$$

But

$$\sum_c \mathrm{Spur}(W_c^d)\, \mathrm{Spur}(W_c^\beta) = \delta_{d\beta}\, h,$$

where h = order of the group. Hence

$$n_d = \tfrac{1}{h} \sum_c \mathrm{Spur}(\mathcal{O}_c)\, \mathrm{Spur}(W_c^d). \tag{39}$$

Now both $\mathrm{Spur}(W_c^d)$ and d_α can be found[3] from a table of characters of the point group, and $\mathrm{Spur}(\mathcal{O}_c)$ can be calculated through the theorem in §5, so that d_α and n_α are easily calculable from (39).

EXAMPLE We consider again the molecule CH_3Cl. (§5) From the left table of characters we get

Representation	d_α	Spur(W_c) $c=C_1$	$c=C_2, C_3, C_4$	$c=C_5, C_6$
W^1	1	1	1	1
W^2	1	1	1	-1
W^3	2	2	-1	0

$n_1 = 3$, $n_2 = 0$, $n_3 = 3$.

It follows that the secular equ. is factored into 3 cubic equs. 2 of which are identical.

We conclude that there are totally 6 vibrational frequencies 3 of which are doubly degenerate.

The no. of independent constants in the potential energy is (§9) $\sum_\alpha n_\alpha^2 = 18$.

Other Developments

§18 ISOTOPE RULE[7] In the above example the no. of unknown constants in the potential energy is much greater than the no. of observable frequencies, as is usually the case. The situation is, however, not so hopeless, because the frequencies of the molecule with some of the atoms replaced by their isotopes serve as additional data. The new molecule has evidently the same equilibrium configuration and potential energy as the original one. But in calculating its kinetic energy it must be remembered that all the results of §§1 to 7 are correct, (because they do not concern the masses of the nuclei) but that since now $\left(\frac{m_i}{m_N}\right)$ does not necessarily commute with P_c, (21) does not hold. The secular equ. is therefore not factored as before. The vectorial method is still convenient.

If the new kinetic energy matrix is denoted by g', from (34)

$$\pi \lambda_i = |\mathcal{H} g| = |\mathcal{H}||g|,$$
$$\pi \lambda_i' = |\mathcal{H} g'| = |\mathcal{H}||g'|.$$

Hence $\pi\lambda_i / \pi\lambda_i' = |g|/|g'| = \left|\mathcal{L}\left(\frac{m_i}{m_N}\right)\mathcal{L}'\right| / \left|\mathcal{L}\left(\frac{m_i'}{m_N'}\right)\mathcal{L}'\right|$.

§19 SPLITTING OF HIGH FREQUENCIES[7] Sometimes it is known that certain force constants are considerably larger than the others, so that the corresponding frequencies of vibration are much greater than the rest. To solve for the low frequencies we may make the approximation of putting the large force constants equal to infinity. Let

$$g = \begin{pmatrix} g_{11} & g_{12} \\ g_{21} & g_{22} \end{pmatrix}, \quad \mathcal{H} = \begin{pmatrix} \mathcal{H}_{11} & \mathcal{H}_{12} \\ \mathcal{H}_{21} & \mathcal{H}_{22} \end{pmatrix}.$$

We want to find the λ's when $\mathcal{H}_{11} \longrightarrow \infty$. Now

$$|\lambda g^{-1} \mathcal{H}^{-1} - I| = 0.$$

and when $\mathcal{H}_{11} \to \infty$, $\mathcal{H}^{-1} \to \begin{pmatrix} 0 & 0 \\ 0 & \mathcal{H}_{22}^{-1} \end{pmatrix}$. Since

we get $g^{-1} = \begin{pmatrix} (g^{-1})_{11} & (g^{-1})_{12} \\ (g^{-1})_{21} & (g_{22} - g_{21}g_{11}^{-1}g_{12})^{-1} \end{pmatrix}$,

or $\left|\lambda \begin{pmatrix} 0 & (g^{-1})_{12}\mathcal{H}_{22}^{-1} \\ 0 & (g_{22} - g_{21}g_{11}^{-1}g_{12})^{-1}\mathcal{H}_{22}^{-1} \end{pmatrix} - I\right| = 0,$

i.e. $|\lambda (g_{22} - g_{21}g_{11}^{-1}g_{12})^{-1}\mathcal{H}_{22}^{-1} - I| = 0.$

$|\lambda (g_{22} - g_{21}g_{11}^{-1}g_{12})^{-1} - \mathcal{H}_{22}| = 0.$

-14-
Acknowledgement

The author wishes to express his thanks to Prof. T. Y. Wu for his continual guidance.

++++++++++++++

References

1. von H. Bethe, Ann. d. Physik, 3(1929), 133.
2. E. Wigner, Göttingen Nachrichten, (1930), p.133.
3. See 1. and 2., also Speiser: Theorie der Gruppen von Endlicher Ordnung (Berlin, 1927).
    and J. E. Rosenthal and G. M. Murphy, Rev. Mod. Phys., 8(1936), 317.
    The degeneracy and other properties are tabulated by
    E. B. Wilson Jr. in, J. Chem. Phys., 2(1934), 432.
    (complete to molecules containing 7 atoms)
    The irreducible representations are given in
    F. Seitz, Zeits. f. Krist., A88(1934), 433.
4. E. Wigner, Gruppentheorie .(Braunschweig 1931) This book is always meant when Wigner is referred to.
5. Wigner, p.86.
6. ibid. p.83.
7. E. B. Wilson Jr., J. Chem. Phys., 9(1941), 76.
8. J. E. Rosenthal and G. M. Murphy, Rev. Mod. Phys., 8(1936), 317.
9. E. B. Wilson Jr. and B. L. Crawford Jr., J. Chem. Phys.,8(1936), 223.
10. D. M. Dennison, Rev. Mod. Phys., 3(1931), 280.
11. J. B. Howard and E. B. Wilson, Jr., J. Chem. Phys., 2(1934), 630.

Investigations in the Statistical Theory of Superlattices

1944 thesis for MSc degree

National Tsing Hua University

INVESTIGATIONS IN THE STATISTICAL THEORY OF SUPERLATTICES

A Dissertation
Submitted to
The Faculty of the Graduate School of Science
in Candidacy for
The Degree of Master of Science

By   C. N. Yang   (楊振寧)

Kunming, China
June, 1944

CONTENTS

I. The Variation of the Interaction Energy with Change of Lattice Constants and Change of the Degree of Order

II. A Generalization of the Quasi-chemical Method in the Statistical Theory of Superlattices

THE VARIATION OF THE INTERACTION ENERGY WITH CHANGE OF LATTICE CONSTANTS AND CHANGE OF THE DEGREE OF ORDER

By C. N. Yang

National Tsing Hua University,

Kunming, China

ABSTRACT

The change of the lattice constants due to the order-disordering process in a superlattice is investigated by using the condition of minimum free energy in Bethe's theory. It is found that the interaction energy depends on the degree of order when the external pressure is kept constant. The specific heat at constant pressure given by the theory is compared with experiment. Another source of the variation of interaction energy is the change of atomic arrangements. This is also investigated from the view point of Wang's formulation of the free energy in Bethe's approximations.

1. INTRODUCTION

The binary alloy CuAu is face-centred cubic when disordered and tetragonal when ordered. This change of lattice form can be studied thermodynamically if we know the energy and the entropy of the crystal. Some calculations along this line has already been made by Wilson[1] who used Bethe's method to find the energy but Bragg-Williams' method to find the entropy of the crystal. It will be shown in the present paper that Bethe's method can be carried through in the calculations, making it self-consistent. The results are comparable with Gorsky's measurements[2].

The change of lattice constants evidently affects the interaction energy between the atoms, and must consequently produce a change in the configurational energy and the specific heat of the crystal. We shall see that the effect is in the right direction to bring the theory into closer agreement with experiment, because it tends to make the energy increase more rapidly near the critical temperature. An actual calculation of the specific heat at variable lattice constant but constant external pressure for β-brass is given in section 3.

Now the interaction energy can also be influenced by a change of the atomic arrangements. Mott[3] has shown from a study of the electronic distribution in superlattices that the interaction energy decreases as the degree of order decreases. The actual relation between the two is naturally very complicated. A linear dependence (of the average interaction energy upon the degree of order) has been assumed by Lin[4] in attempting to explain the occurrence of the maximum critical temperature of a face-centred alloy at the concentration ratio 1:3. In order to justify the assumption we shall view the problem from a new angle by the introduction of the free energy in Bethe's approximation[5]. In this way it is found that the interaction energy as a function of the degree of order must satisfy certain equations obtained from a set of conditions of consistency. This same set of conditions of consistency makes also possible the calculation of the energy of the crystal without appealing to Bragg-Williams' theory as Lin did.

2. THE VARIATION OF LATTICE CONSTANTS

We shall form the partition function at constant lattice constants l_1 and l_2, and then obtain their equilibrium values from the equations determining the generalized reactions. Let $\frac{1}{2}zNm$ be the number of A-B neighbors in the crystal. If $g(m)$ is the

---

[1] Wilson, Proc. Camb. Phil. Soc. 34, 81(1938).
[2] Gorsky, Zeit. f. Phys. 50, 64(1928).
[3] Mott, Proc. Phys. Soc. 49, 258(1937).
[4] Lin, Chinese J. Phys. 3, 182(1939).
[5] Wang,"Free Energy in the Statistical Theory of Order-Disorder Transformations", Science Report of National Tsing Hua University, series A, 30-th anniversary Memorial Number (1941), printed but failed to appear.

number of arrangements of the atoms for the given value of m, and $W(l_1, l_2, m)$ the configurational energy of the crystal, the configurational partition function is

$$f(m, T, l_1, l_2) = g(m) \exp(-W/kT).$$

The equilibrium value $\bar{m}$ of m is determined from the condition of a maximum of f:

$$\frac{\partial}{\partial \bar{m}} \log f(\bar{m}, T, l_1, l_2) = 0.$$

The generalized reactions are given by

$$L_i = kT\frac{\mathrm{d}}{\mathrm{d}l_i}\log f(\bar{m}, T, l_1, l_2) = kT\frac{\mathrm{d}\bar{m}}{\mathrm{d}l_i}\frac{\partial}{\partial \bar{m}}\log f + kT\frac{\partial}{\partial l_i}\log f = kT\frac{\partial}{\partial l_i}\log f$$
$$= -\frac{\partial}{\partial l_i}W(l_1, l_2, \bar{m}). \tag{1}$$

To study the change of lattice form in CuAu we divide the face-centred lattice into four simple cubic sublattices 1, 2, 3, 4[*]. Let the shortest distance between the sites of 1 and 2, or 3 and 4 be l_1, that between the sites of 1 and 3, 1 and 4, 2 and 3 or 2 and 4 be l_2, so that the former is the distance between neighbouring Au-Au or Cu-Cu atoms and the latter that between neighbouring Au-Cu atoms when the crystal is perfectly ordered. The interaction energies V_{AA}, V_{AB} and V_{BB} are functions of l_1, and l_2.

If the number of sites of each sublattice is $\frac{1}{2}N$, the number of pairs of sites between sublattices 1 and 2 must be $4(\frac{1}{2})N = 2N$. Denote by m_{12} the fraction of A-B pairs among these. Then the number of

A-A pairs is $\quad \frac{1}{2}\left[4\left(\frac{N}{2}\theta_1 + \frac{N}{2}\theta_2\right) - 2Nm_{12}\right] = N[\theta_1 + \theta_2 - m_{12}],$

B-B pairs is $\quad \frac{1}{2}\left[4\left(\frac{N}{2}\{1-\theta_1\} + \frac{N}{2}\{1-\theta_2\}\right) - 2Nm_{12}\right] = N[2 - \theta_1 - \theta_2 - m_{12}],$

where θ_i is the fraction of sites of sublattice i occupied by A atoms. Thus the energy of interaction between the atoms on sublattices 1 and 2 is

$$N[(\theta_1 + \theta_2 - m_{12})V_{AA}(l_1) + 2m_{12}V_{AB}(l_1) + (2 - \theta_1 - \theta_2 - m_{12})V_{BB}(l_1)].$$

Writing

$$c = \frac{1}{4}(\theta_1 + \theta_2 + \theta_3 + \theta_4) \quad \text{and} \quad V = \frac{1}{2}(V_{AA} + V_{BB}) - V_{AB},$$

we get the energy of the whole crystal

$$W = N[4cV_{AA}(l_1) + 4(1-c)V_{BB}(l_1) - 2(m_{12} + m_{34})V(l_1)] +$$
$$N[8cV_{AA}(l_2) + 8(1-c)V_{BB}(l_2) - 2(m_{13} + m_{14} + m_{23} + m_{24})V(l_2)]. \tag{2}$$

---

[*]Cf. Fig.27 in Rev. Mod. Phys. 10, 1(1938).

With this value for W, (1) becomes
$$L_1 = -N[4cV'_{AA}(l_1) + 4(1-c)V'_{BB}(l_1) - 2(\bar{m}_{12} + \bar{m}_{34})V'(l_1)], \quad (3)$$
and
$$L_2 = -N[8cV'_{AA}(l_2) + 8(1-c)V'_{BB}(l_2) - 2(\bar{m}_{13} + \bar{m}_{14} + \bar{m}_{23} + \bar{m}_{24})V'(l_2)]. \quad (4)$$

To solve for l_1 and l_2 as functions of T we must first know the $\bar{m}$'s, which are usually very complicated. Wilson[1] discussed the values of l_1 and l_2 only in the cases when the alloy is disordered and when the order is nearly perfect. We shall also confine our attention to these cases.

(i) <u>Disordered</u>. In this case there is no difference between the four sublattices so that all the $\bar{m}_{ij}$'s are equal to $\bar{m}$. (3) and (4) reduce to
$$\begin{aligned} L_1 &= -N[4cV'_{AA}(l_1) + 4(1-c)V'_{BB}(l_1) - 4\bar{m}V'(l_1)], \\ L_2 &= -N[8cV'_{AA}(l_2) + 8(1-c)V'_{BB}(l_2) - 8\bar{m}V'(l_2)]. \end{aligned} \quad (5)$$

If $L_1 = L_2 = 0$, this shows that $l_1 = l_2$, so that the crystal is cubic.

(ii) <u>Order nearly perfect</u>. When $c = \frac{1}{2}$, and the order is nearly perfect,
$$\theta_1 = \theta_2 \cong 1, \quad \theta_3 = \theta_4 \cong 0, \quad \theta_1 + \theta_3 = 1, \quad \theta_1 - \theta_3 = s.$$

There are only a few B atoms on sublattices 1 and 2. Hence approximately
$$\bar{m}_{12} = (1 - \theta_1) + (1 - \theta_2) = 2\theta_3 = 1 - s.$$

By the same reason we can obtain the number of A-A pairs of neighbours between the sublattices 1 and 3:
$$N(\theta_1 + \theta_3 - \bar{m}_{13}) = 4\left(\frac{1}{2}N\theta_3\right).$$

Thus
$$\bar{m}_{13} = \theta_1 - \theta_3 = s.$$

We can now write down all the $\bar{m}$'s:
$$\bar{m}_{12} = \bar{m}_{34} = 1 - s, \quad \bar{m}_{13} = \bar{m}_{23} = \bar{m}_{14} = \bar{m}_{24} = s.$$

These equations are correct to the first order of $(1-s)$. Substituting them into (3) and (4) we obtain
$$\begin{aligned} L_1 &= -2N[V'_{AA}(l_1) + V'_{BB}(l_1) - 2(1-s)V'(l)], \\ L_2 &= -4N[V'_{AA}(l_2) + V'_{BB}(l_2) - 2sV'(l_2)]. \end{aligned}$$

These are exactly equations (27) in Wilson's paper, from which an expression of the degree of tetragonality in agreement with Gorsky's measurements[2] can be obtained.

3. THE EFFECT OF THE CHANGE OF LATTICE CONSTANTS ON THE INTERACTION ENERGY

In the alloy CuAu the gold atoms and the copper atoms are in contact when the order is perfect. Since the copper atom is somewhat smaller than the gold atom, the size of the crystal must increase when gold atoms exchange their positions with copper atoms. Thus with increasing disorder the distance between the atoms increases and hence the interaction energies diminish. The disordering process is therefore effected with more ease near the critical temperature than it is at lower temperatures; and we expect the specific heat at constant pressure to possess a steeper and higher maximum at the critical temperature than the specific heat at constant volume.

Now we shall calculate in length the specific heat at constant pressure of the alloy β-brass, which forms the simplest type of superlattice that can be studied statistically. Bethe's method will be used.

The configurational energy of the crystal is, in Easthope's[6] notations:

$$W = -N_{AB}V + \frac{1}{2}Nz[c(V_{AA} - V_{BB}) + V_{BB}]. \tag{6}$$

Substitution of this expression into (1) gives

$$0 = -\bar{m}V'(l) + c[V'_{AA}(l) - V'_{BB}(l)] + V'_{BB}(l),$$

when the pressure is put equal to zero. Now the variation of V is not very large, so that to a sufficient approximation we may assume the linear relations

$$[V'_{AA}(l) - V'_{BB}(l)]/V'(l) = -K_0 + K_1 V, \tag{7}$$

and

$$V'_{BB}(l)/V'(l) = -J_0 + J_1 V. \tag{8}$$

These three last equations give, after eliminating $V'_{AA}(l)$ and $V'_{BB}(l)$:

$$V = \frac{\bar{m} + (cK_0 + J_0)}{cK_1 + J_1}. \tag{9}$$

We have already seen that V increases as $\bar{m}$ increases, hence $cK_1 + J_1$ must be positive. The other constant $cK_0 + J_0$ must also be positive in order that V may be positive with only a relatively small variation.

Eisenschitz[7] has calculated the specific heat at constant pressure by Bragg-Williams' method. He assumed that the interaction energies depend on a parameter u in the following way:

$$\frac{1}{2}(V_{AA} + V_{BB}) = \phi[(1-a) + a(1-u)^2], \qquad V_{AB} = \phi bu^2,$$

---
[6] Easthope, Proc. Camb. Phil. Soc. 33, 502(1937); 34, 68(1938).
[7] Eisenschitz, Proc. Roy. Soc. 68, 546(1938).

where $a = .225$, $b = .203$ and u is of the order of unity. Comparing this with (7) and (8) we see that his assumption is equivalent (approximately) to ours if $\frac{1}{2}K_0 + J_0 = 1.22$ and $\frac{1}{2}K_1 + J_1 = .508 \times 10^{14} \text{erg}^{-1}$. But with these values the specific heat at the critical temperature would be too large. In order to make $(C_p)_{T_c} = 5.1R$ as given by the measurements of Sykes and Wilkinson[8] we assume (cf. eq.(12) below)

$$\frac{1}{2}K_0 + J_0 = 1.79.$$

With this value for $\frac{1}{2}K_0 + J_0$, the relative variation of V can be shown to be within 1.3%.

We can now start from (9) and the equations given by Easthope[6] for the determination of $\bar{m}$ as a function of the temperature and V to obtain the specific heat at constant pressure:

$$C_p = \frac{dW}{dT} = \left(\frac{\partial W}{\partial \bar{m}}\right)_l \frac{d\bar{m}}{dx}\frac{dx}{dT} + \left(\frac{\partial W}{\partial l}\right)_{\bar{m}} \frac{dl}{dT} = \left(\frac{\partial W}{\partial \bar{m}}\right)_l \frac{d\bar{m}}{\partial x}\frac{dx}{dT}. \tag{10}$$

But

$$\frac{dx}{dT} = \frac{xV/kT^2}{1 + \frac{x}{kT}\frac{dV}{d\bar{m}}\frac{d\bar{m}}{dx}}. \tag{11}$$

Hence by (6)

$$C_p = \frac{\frac{1}{2}Nzkx(\log x)^2(-\frac{d\bar{m}}{dx})}{1 + \frac{x\log x}{\bar{m}+(cK_0+J_0)}(-\frac{d\bar{m}}{dx})}. \tag{12}$$

The value of this expression is calculated for the case $c = \frac{1}{2}$, the constant $\frac{1}{2}K_0 + J_0$ being assumed to be 1.79 to make $(C_p)_{T_c} = 5.1R$. The result is plotted in the accompanying figure together with Bethe's curve[9] for C_V and Sykes ad Wilkinson's experimental[8] data.

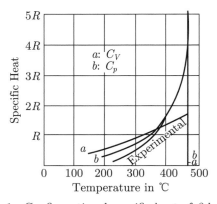

Fig. 1 Configurational specific heat of β-brass.

---

[8] Sykes and Wilkinson, Inst. Metals J. 61, 223(1937).
[9] Nix and Shockley, Rev. Mod. Phys. 10, 1(1938).

4. THE EFFECT OF THE ATOMIC DISTRIBUTION ON THE INTERACTION ENERGY

As has already been mentioned, the interaction energy depends in some very complicated manner upon the degree of order. To study the effect of such a dependence Lin[4] has assumed a linear relationship:

$$V = V_0(1 + \alpha c + \beta m_{AA}) \tag{13}$$

between the interaction energy: V and the fraction of A-A pairs of neighbours: m_{AA}. In this section we shall study the general nature of the variation of V in the light of the theory of the free energy in Bethe's approximation given by Wang[5].

The fundamental equations in Wang's paper are (45), (46) and (39) with ξ_α and ξ_β given by (47), (48), (49) and (50). These equations are still assumed to be valid now V becomes a function of θ_α, θ_β and T. They may be put into the form:

$$\frac{d}{d\theta_\alpha} \log Q = -Nr_\alpha \log \xi_\alpha = \frac{\partial}{\partial \theta_\alpha} \log Q_0(\theta_\alpha, \theta_\beta, V/T), \tag{14}$$

$$\frac{d}{d\theta_\beta} \log Q = -Nr_\beta \log \xi_\beta = \frac{\partial}{\partial \theta_\beta} \log Q_0(\theta_\alpha, \theta_\beta, V/T), \tag{15}$$

$$E = kT^2 \frac{d}{dT} \log Q, \tag{16}$$

where Q_0 is the partition function for the case when V is a constant, if we denote by $\frac{\partial}{\partial \theta_\alpha}$, $\frac{\partial}{\partial \theta_\beta}$, $\frac{\partial}{\partial T}$ and $\frac{\partial}{\partial V}$ differentiations when θ_α, θ_β, T and V are regarded as independent of each other, and by $\frac{d}{d\theta_\alpha}$ the operator $\frac{\partial}{\partial \theta_\alpha} + \frac{dV}{d\theta_\alpha} \frac{\partial}{\partial V}$. (14) and (15) mean that we have assumed with Lin that the equilibrium values of θ_α and θ_β are given by the same equations as in Bethe's approximation. (16) gives the energy of the crystal.

Consistency of (14) and (15) requires

$$\frac{d}{d\theta_\alpha}\left(\frac{\partial}{\partial \theta_\beta} \log Q_0\right) = \frac{d}{d\theta_\beta}\left(\frac{\partial}{\partial \theta_\alpha} \log Q_0\right),$$

i.e.

$$\frac{dV}{d\theta_\alpha} \frac{\partial^2}{\partial \theta_\beta \partial V} \log Q_0 = \frac{dV}{d\theta_\beta} \frac{\partial^2}{\partial V \partial \theta_\alpha} \log Q_0. \tag{17}$$

Upon the hypothesis of nearest neighbour interaction the energy in Wang's paper becomes

$$E_0 = kT^2 \frac{\partial}{\partial T} \log Q_0 = zNV m_{AA}.$$

But Q_0 depends on T and V through V/T, so that

$$\frac{\partial^2}{\partial \theta_\alpha \partial V} \log Q_0 = -\frac{T}{V} \frac{\partial^2}{\partial T \partial \theta_\alpha} \log Q_0 = -\frac{\partial E_0}{\partial \theta_\alpha}/kTV.$$

Substituting this eq. and a similar one into (17) we obtain

$$\frac{\partial E_0}{\partial \theta_\beta}\frac{\mathrm{d}V}{\mathrm{d}\theta_\alpha} = \frac{\partial E_0}{\partial \theta_\alpha}\frac{\mathrm{d}V}{\mathrm{d}\theta_\beta}, \tag{18}$$

which becomes, if $\frac{\partial E_0}{\partial V}\frac{\mathrm{d}V}{\mathrm{d}\theta_\alpha}\frac{\mathrm{d}V}{\mathrm{d}\theta_\beta}$ is added to both sides,

$$\frac{\mathrm{d}E_0}{\mathrm{d}\theta_\alpha}\frac{\mathrm{d}V}{\mathrm{d}\theta_\beta} = \frac{\mathrm{d}E_0}{\mathrm{d}\theta_\beta}\frac{\mathrm{d}V}{\mathrm{d}\theta_\alpha}.$$

This shows that V and E_0 are connected by a relation independent of θ_α and θ_β, i.e.

$$V = V(E_0, T). \tag{19}$$

A consequence of this result is that the coefficient α in Lin's relation (13) must be zero. This makes, however, the maximum critical temperature for the AB type of superlattice to shift to a value of the concentration different from $\frac{1}{2}$, which contradicts experimental results. One way out of the difficulty is to make some other assumption regarding the dependence of V on the degree of order, such as

$$V = V_0\left[1 + \beta m_{AA} - \frac{\beta(z-1)}{2c(cz-1)}m_{AA}^2\right]. \tag{20}$$

Let us now try to find E in the general case. From (14) and (16) we get

$$\frac{\mathrm{d}E}{\mathrm{d}\theta_\alpha} = kT^2\frac{\mathrm{d}}{\mathrm{d}T}\left(\frac{\partial}{\partial\theta_\alpha}\log Q_0\right) = kT^2\frac{\mathrm{d}(\frac{V}{T})}{\mathrm{d}T}\frac{\partial^2}{\partial\theta_\alpha\partial(\frac{V}{T})}\log Q_0 = -\frac{T^2}{V}\frac{\mathrm{d}(\frac{V}{T})}{\mathrm{d}T}\frac{\partial E_0}{\partial\theta_\alpha}. \tag{21}$$

Similarly

$$\frac{\mathrm{d}E}{\mathrm{d}\theta_\beta} = -\frac{T^2}{V}\frac{\mathrm{d}(\frac{V}{T})}{\mathrm{d}T}\frac{\partial E_0}{\partial\theta_\beta}.$$

Hence

$$\frac{\mathrm{d}E}{\mathrm{d}\theta_\alpha}\frac{\partial E_0}{\partial\theta_\beta} = \frac{\partial E_0}{\partial\theta_\alpha}\frac{\mathrm{d}E}{\mathrm{d}\theta_\beta}.$$

Just as (18) leads to (19), this last equation leads to

$$E = E(E_0, T).$$

Substituting this into (21) we obtain

$$\frac{\partial E}{\partial E_0}\left(\frac{\partial E_0}{\partial\theta_\alpha} + \frac{\partial E_0}{\partial V}\frac{\partial V}{\partial\theta_\alpha}\right) = \frac{\partial E_0}{\partial\theta_\alpha}\left(1 - \frac{T}{V}\frac{\mathrm{d}V}{\mathrm{d}T}\right).$$

But (19) gives

$$\frac{\mathrm{d}V}{\mathrm{d}\theta_\alpha} = \frac{\frac{\partial E}{\partial\theta_\alpha}\frac{\partial V}{\partial E_0}}{1 - \frac{\partial V}{\partial E_0}\frac{\partial E_0}{\partial V}} \quad \text{and} \quad \frac{\mathrm{d}V}{\mathrm{d}T} = \frac{\frac{\partial V}{\partial T} + \frac{\partial V}{\partial E_0}\frac{\partial E_0}{\partial T}}{1 - \frac{\partial V}{\partial E_0}\frac{\partial E_0}{\partial V}}.$$

Hence
$$\frac{\partial E}{\partial E_0} = \left(1 - \frac{\partial V}{\partial E_0}\frac{\partial E_0}{\partial V}\right)\left(1 - \frac{T}{V}\frac{dV}{dT}\right) = 1 - \frac{\partial V}{\partial E_0}\frac{\partial E_0}{\partial V} - \frac{T}{V}\frac{dV}{dT} - \frac{T}{V}\frac{\partial V}{\partial E_0}\frac{\partial E_0}{\partial T}.$$

Now E_0/V is a function of θ_α, θ_β and V/T, so that
$$\frac{\partial E_0}{\partial V} = -\frac{T}{V}\frac{\partial E_0}{\partial T} + \frac{\partial E_0}{\partial V}.$$

Hence
$$\frac{\partial E}{\partial E_0} = 1 - \frac{E_0}{V}\frac{\partial V}{\partial E_0} - \frac{T}{V}\frac{\partial V}{\partial T}. \tag{22}$$

If V depends on m_{AA} only, and not on T, we have
$$\frac{\partial E}{\partial E_0} = 1 - \frac{E_0}{V}\frac{dV}{dE_0} = V\frac{d(E_0/V)}{dE_0}.$$

The boundary condition is given by the case when there is no A atoms, i.e. when $\theta_\alpha + \theta_\beta = 0$. In this case $m_{AA} = 0$, $E_0 = E = 0$. Hence
$$E = \int_0^{E_0}\left[V\frac{d(E_0/V)}{dE_0}\right]dE_0 = zN\int_0^{m_{AA}}V\,dm_{AA}. \tag{23}$$

If the assumption (20) is made, the energy of the crystal is
$$E = zNV_0\left[m_{AA} + \frac{\beta}{2}m_{AA}^2 - \frac{\beta(z-1)}{6c(cz-1)}m_{AA}^3\right].$$

When β is not large, this differs very little in numerical value from Bethe's original expression. The specific heat is
$$C_V = zNV\frac{dm_{AA}}{dT} = zNV\frac{dm_{AA}}{dx}\frac{\frac{xV}{kT^2}}{1 + \frac{x}{kT}\frac{dV}{dx}} = zNkx(\log x)^2\frac{dm_{AA}}{dx}\frac{1}{1 + \frac{x}{kT}\frac{dV}{dm_{AA}}\frac{dm_{AA}}{dx}}.$$

The author is very much indebted to Prof. J. S. Wang for suggesting this problem and for helpful discussions.

A GENERALIZATION OF THE QUASI-CHEMICAL METHOD IN THE STATISTICAL THEORY OF SUPERLATTICES

By C. N. Yang[*]

National Tsing Hua University,

Kunming, China

ABSTRACT

The quasi-chemical method introduced by Fowler and Guggenheim for the equilibrium distribution of pairs of sites in a superlattice is generalized. It is shown that by considering groups containing large numbers of sites the method may be used to obtain successive approximations of the free energy of the crystal. To analyze the fundamental assumption underlying the method more closely, the hypothesis of the non-interference of local configurations is discussed. The free energy of the crystal is obtained without integration as a closed algebraic expression with the aid of a Legendre transformation. Applications of the results are then made to different approximations for simple and body-centred cubic crystals and for the face-centred cubic crystal Cu_3Au. In each case the free energy is obtained and discussed.

---

[*]Research Fellow of the China Foundation for the Promotion of Education and Culture.

1. INTRODUCTION

It was shown by Fowler and Guggenheim[1] that the quasi-chemical method, originally devised for the theory of regular solutions, applies equally well to the theory of superlattices with long-distance order. The method is, as they have emphasized, definitely one stage further towards an exact theory than Bragg-Williams' method[2]. When compared with Bethe's[3] or Kirkwood's[4] method it also distinguishes itself in mathematical simplicity. But to be a method that may lead to a consistent scheme of successive approximations, it must be applicable to the n-th approximation in the theory of superlattices. This does not, however, seem possible in the original form of the method given by Fowler and Guggenheim. It is the purpose of the present paper to formulate the quasi-chemical method in a new way which is applicable to high order approximations in the theory of superlattices.

The free energy expression in Bethe's and in the quasi-chemical methods involves an integral. Its evaluation is very complicated and has been carried out[1,5] so far only in Bethe's approximation for simple and body-centered cubic crystals. In the new formulation of the quasi-chemical method, however, it will be shown that a Legendre transformation helps much in avoiding the mathematical difficulties. (It might be noticed that a similar Legendre transformation is used to essentially the same effect in Fowler's formulation of general statistical mechanics. Cf. Fowler, Statistical Mechanics, second edition, p.188.) The free energy is obtained directly as a closed algebraic expression. Its values are given for Bethe's first and second (modified) approximations and for the face-centered alloy Cu_3Au in sections 7 and 8.

To make sure that the quasi-chemical method may actually be used to obtain a series of successively better approximations, we must investigate the free energy in high order approximations and compare it with the partition function of the crystal. This is done in section 5 together with a comparison of the quasi-chemical and Bethe's methods.

Except in the last section we are only concerned with binary alloys with atomic ratio 1:1 forming a (quadratic), simple cubic or body-centered cubic lattice. The generalization of the method to the investigation of alloys with other atomic ratios and forming other types of lattices is easy. In fact, the superior power of the quasi-

---

[1] Fowler and Guggenheim, Proc. Roy. Soc. A174, 189 (1940).
[2] Bragg and Williams, Proc. Roy. Soc. A145, 699 (1934); 151, 540 (1935); 152, 231 (1935).
[3] Bethe, Proc. Roy. Soc. A150, 552 (1935).
[4] Kirkwood, J. Chem. Phys. 6, 70 (1938).
[5] Chang, Proc. Camb. Phil. Soc. 35, 265 (1939); Kirkwood, J. Chem. Phys. 8, 623 (1940); Wang, "Free Energy in the Statistical Theory of Order–Disorder Transformation", Science Report of National Tsing Hua University, Series A, 30-th Anniversary Memorial Number (1941), printed but failed to appear.

chemical method appears to be even more fully revealed when a face-centered lattice is treated. This problem is taken at the end of the paper where an approximate free energy expression for Cu$_3$Au is obtained and its critical phenomena discussed.

2. REFORMULATION OF THE QUASI-CHEMICAL METHOD

Consider a crystal AB. Let $2N$ be the total number of atoms and z the number of nearest neighbours of each. At low temperatures we can distinguish between the so-called α-sites and β-sites for A and B atoms respectively. Denote by Nr the number of A atoms on α-sites. Let

$$1 - r = w, \qquad r - w = s.$$

The partition function of the crystal is

$$\sum_r p(r, T), \tag{1}$$

where $p(r, T)$ is equal to $\sum \exp(-E/kT)$ over all possible configurations of the crystal with the given value of r. The average energy over all these configurations is

$$\overline{E}(r, T) = kT^2 \frac{\partial}{\partial T} \log p(r, T). \tag{2}$$

But evidently,

$$p(r, \infty) = g(r) = \left[\frac{N!}{(Nr)!(N\omega)!}\right]^2. \tag{3}$$

Hence

$$\log p(r, T) = \log g(r) + \int_\infty^T \frac{1}{kT^2} \overline{E}(r, T) \mathrm{d}T. \tag{4}$$

The problem therefore reduces to one of finding $\overline{E}(r, T)$. Since a direct solution is very difficult we shall try to find an approximate solution by the quasi-chemical method, which is presented below in a form easily generalizable.

There are in the crystal zN nearest pairs of sites α-β. Among these let there be $[q_\alpha, q_\beta]$ with their α-sites occupied by q_α ($= 0, 1$) wrong (B) atoms and their β-sites by q_β ($= 0, 1$) wrong (A) atoms. For given r the following relations hold:

$$[0, 1] + [0, 0] + [1, 1] + [1, 0] = zN,$$
$$[1, 1] + [1, 0] = zNw, \tag{5}$$
$$[0, 1] \quad + [1, 1] \quad = zNw.$$

Upon the approximation of neighbour interaction the energy of the crystal may be written as

$$E(r, T) = [0, 1]V_{AA} + [0, 0]V_{AB} + [1, 1]V_{AB} + [1, 0]V_{BB}, \tag{6}$$

where the V's are the interaction energies between a pair of nearest neighbours.

We may give (5) and (6) a different interpretation by imagining [0,1], [0,0], [1,1], [1,0] and V_{AA}, V_{AB}, V_{AB}, V_{BB} to be respectively the numbers and the molecular internal energies of the four different kinds of molecules XZ, X, XYZ, XY of a gaseous assembly. The interpretation of (6) is that the assembly has the same internal (non-kinetic) energy as the crystal at the given value of r. (5) would mean that there are altogether zN X atoms, zNw Y atoms and zNw Z atoms in the assembly.

The quasi-chemical method consists in taking the averages $\overline{[0,1]}, \overline{[0,0]}, \overline{[1,1]}, \overline{[1,0]}$ of the assembly at any temperature as approximately representing the corresponding averages of the crystal at the same temperature. Whether this approximation is good can only be judged for the present from the results it leads to.

A detailed treatment of the problem of a gaseous assembly has been given by Fowler[6]. We are only interested in our assembly of four different kinds of molecules, for which the results may be summarized as:

$$\overline{[0,1]} = \xi\nu e^{-V_{AA}/kT}, \qquad \overline{[0,0]} = \xi e^{-V_{AB}/kT}, \qquad (7)$$
$$\overline{[1,1]} = \xi\mu\nu e^{-V_{AB}/kT}, \qquad \overline{[1,0]} = \xi\mu e^{-V_{BB}/kT},$$

where ξ, μ and ν are to be determined from (6). From (7) we get

$$\frac{\overline{[0,0]}\,\overline{[1,1]}}{\overline{[0,1]}\,\overline{[1,0]}} = x^{-2}, \qquad (8)$$

where

$$x = \exp\left[-\frac{1}{2}(V_{AA} + V_{BB} - 2V_{AB})/kT\right]. \qquad (9)$$

(8) and (6) together form the starting point of Fowler and Guggenheim's work[1]. The subsequent calculations of $\overline{E}(r,T)$, $p(r,T)$ and the free energy of the crystal are straightforward and will not be repeated here. We shall see later how the free energy can be written down directly without actual integration.

3. GENERALIZATION TO GROUPS OF FOUR SITES

So far we have fixed our attention on the pairs of nearest neighbours in the crystal and have used the quasi-chemical method to obtain the average numbers of the four different kinds of pairs. Now we shall generalize the whole procedure: we shall study all the groups of sites of an arbitrarily chosen form in the crystal, and classifying these groups according to the way they are occupied by atoms we shall obtain the average number of groups in each class by chemical analogy.

---

[6]Fowler, Statistical Mechanics, second edition, pp. 162–163.

$$
\begin{array}{ccccc}
\beta & \alpha & \beta & \alpha & \beta \\
\alpha & \boxed{\beta \quad \alpha} & \beta & \alpha \\
\beta & \boxed{\alpha \quad \beta} & \alpha & \beta \\
\alpha & \beta & \alpha & \beta & \alpha
\end{array}
$$

Fig. 1

To make this clear let us consider in detail groups of four sites forming squares (as shown) in a quadratic lattice. We classify these groups into $2^4 = 16$ classes denoted by $(0,0,0,0)$, $(0,0,0,1)$, $\cdots$, $(1,1,1,1)$ respectively, so that all groups in the class (q_1, q_2, q_3, q_4) have q_1 wrong atoms in their upper α-sites, q_2 wrong atoms in their lower α-sites, q_3 wrong atoms in their upper β-sites and q_4 wrong atoms in their lower β-sites. The total number of these groups is N. Hence

$$\sum_{q_i=0}^{1} [q_1, q_2, q_3, q_4] = N, \tag{10}$$

where $[q_1, q_2, q_3, q_4]$ is an abbreviation for the number of groups in the class (q_1, q_2, q_3, q_4). Now the number of all those groups in the crystal with a wrong (B) atom on the upper α-site is just the number of B atoms on the α-sites. Hence

$$\sum_{q} q_i [q_1, q_2, q_3, q_4] = Nw, \qquad i = 1, 2, 3, 4. \tag{11}$$

Let $\chi(q_1, q_2, q_3, q_4)$ be the energy of each group in the class (q_1, q_2, q_3, q_4). It is easy to show that the total energy of the crystal is

$$\overline{E}(r,T) = \sum_{q} \overline{[q_1, q_2, q_3, q_4]} \chi(q_1, q_2, q_3, q_4). \tag{12}$$

We may give (10), (11) and (12) an interpretation similar to the one given in section 2 for equations (5) and (6). The same quasi-chemical method used there to obtain (7) leads now to the following averages (approximate) at a given value of r:

$$\overline{[q_1, q_2, q_3, q_4]} = \xi \mu_1^{q_1} \mu_2^{q_2} \mu_3^{q_3} \mu_4^{q_4} e^{-\chi(q_1,q_2,q_3,q_4)/kT}. \tag{13}$$

In this expression the parameters ξ, μ_1, μ_2, μ_3 and μ_4 are to be determined from (10) and (11), which may be written in the form

$$\xi \frac{\partial \phi}{\partial \xi} = N, \qquad \mu_i \frac{\partial \phi}{\partial \mu_i} = Nw \quad (i = 1, 2, 3, 4), \tag{14}$$

if we put

$$\phi(\xi, \mu_1, \mu_2, \mu_3, \mu_4) = \sum_{q} \xi \mu_1^{q_1} \mu_2^{q_2} \mu_3^{q_3} \mu_4^{q_4} e^{-\chi(q_1,q_2,q_3,q_4)/kT}. \tag{15}$$

Or again, in the form

$$\frac{\partial \Psi}{\partial \log \xi} = \frac{\partial \Psi}{\partial \log \mu_1} = \frac{\partial \Psi}{\partial \log \mu_2} = \frac{\partial \Psi}{\partial \log \mu_3} = \frac{\partial \Psi}{\partial \log \mu_4} = 0, \qquad (16)$$

if we put

$$\Psi = -N \log \xi - \sum_i Nw \log \mu_i + \phi. \qquad (17)$$

It can be shown[7] that ξ and μ_i are uniquely determined by (16) at given r and T. Their values at $T = \infty$ are

$$(\xi)_{T=\infty} = Nr^4, \qquad (\mu_i)_{T=\infty} = \frac{w}{r}, \quad i = 1, 2, 3, 4, \qquad (18)$$

as can be verified by substitution into (14).

To calculate the free energy it is necessary first to evaluate the integral in (4). We shall show that this can be done without first solving (14) for ξ and μ_i. For, by (12) and (13) the integrand may be written

$$\frac{1}{kT^2}\overline{E}(r,T) = \frac{1}{kT^2}\sum_q \overline{[q_1, q_2, q_3, q_4]}\, \chi(q_1, q_2, q_3, q_4) = \frac{\partial \phi}{\partial T}. \qquad (19)$$

In the partial differentiation in $\frac{\partial \phi}{\partial T}$, ξ and μ_i are treated as independent variables. If, however, we regard them as functions (defined by (14)) of r and T, (16) and (17) lead to the following result:

$$\frac{1}{kT^2}\overline{E}(r,T) = \frac{\partial \phi}{\partial T} = \frac{2\Psi(r,T)}{\partial T}. \qquad (20)$$

Mathematically the change of the independent variables from T, ξ and μ_i to T and r is equivalent to the Legendre transformation

$$\xi, \mu_1, \mu_2, \mu_3, \mu_4 \to N, Nw, Nw, Nw, Nw$$

defined by (14). Substituting (20) into (4) we get

$$\log p(r,T) = \log g(r) + \Psi(r,T) - \Psi(r, \infty), \qquad (21)$$

so that the free energy may be written down:

$$F(r,T) = -kT \log p(r,T) = [\log g(r) + \Psi(r,T) - \Psi(r, \infty)](-kT). \qquad (22)$$

The equilibrium value $\bar{r}$ of r is obtained by minimizing F:

$$0 = \frac{\partial F(\bar{r}, T)}{\partial \bar{r}} = -kT\left[\frac{d \log g(\bar{r})}{d\bar{r}} + \frac{\partial \Psi(\bar{r}, T)}{\partial \bar{r}} - \frac{\partial \Psi(\bar{r}, \infty)}{\partial \bar{r}}\right]. \qquad (23)$$

---

[7] The proof follows easily (if we put e^ψ to be the function Φ) from Lemma 2.42 of Fowler's Statistical Mechanics, second edition.

But by (16) and (17)
$$\frac{\partial \Psi(r,T)}{\partial r} = \sum_i N \log \mu_i, \qquad (24)$$
and by (3)
$$\frac{d \log g(r)}{dr} = 2N \log \frac{w}{r},$$
so that by (18)
$$\sum_i \log \mu_i = -2 \log \frac{\overline{w}}{\overline{r}} + \left[\sum_i \log \mu_i\right]_{T=\infty} = 2 \log \left(\frac{\overline{w}}{\overline{r}}\right),$$
i.e.
$$\prod_i \mu_i = \left(\frac{\overline{w}}{\overline{r}}\right)^2. \qquad (25)$$

It will be shown in the next section that we may put $V_{AA} = V_{BB}$, $V_{AB} = 0$, without altering the specific heat of the crystal if $V = \frac{1}{2}(V_{AA} + V_{BB}) - V_{AB}$ is left unchanged. When this is done, ϕ will be symmetrical with respect to μ_1, μ_2, μ_3 and μ_4, and we conclude that all the μ's are equal from the facts that (i) equation (14) has only one set of solution[7], and (ii) if the conclusion is true (14) becomes, with all μ_i put equal to μ,
$$\xi \frac{\partial \phi}{\partial \xi} = N, \qquad \mu \frac{\partial \phi}{\partial \mu} = 4Nw, \qquad (26)$$
which <u>does</u> have[7] a set of solution in ξ and μ. Now ϕ is given by
$$\phi = \xi[1 + 4\mu x^2 + (4\mu^2 x^2 + 2\mu^2 x^4) + 4\mu^3 x^2 + \mu^4], \qquad (27)$$
where x is defined by (9). On eliminating ξ from (26) we obtain
$$(1+s)\mu^4 + (2+4s)x^2\mu^3 + 2sx^2(x^2+2)\mu^2 + (4s-2)x^2\mu + (s-1) = 0. \qquad (28)$$
The free energy is given by (21) and (18):
$$-\frac{F(r,T)}{2NkT} = r \log r + w \log w - 2w \log \mu + \frac{1}{2}\log(1 + 4\mu x^2 + 4\mu^2 x^2 + 2\mu^2 x^4 + 4\mu^3 x^2 + \mu^4), \qquad (29)$$
and the condition of equilibrium by (25):
$$\mu = \sqrt{\frac{1-\overline{s}}{1+\overline{s}}}. \qquad (30)$$

To obtain the critical temperature, we expend (28) in powers of s and find after identifying coefficients
$$\log \mu = -\frac{1 + 6x^2 + x^4}{2 + 2x^2}s + \kappa s^3 + \cdots,$$

which is the only real solution for $\log\mu$. Next we expand (30):

$$\log\mu = -s - \frac{1}{3}s^3 - \cdots .$$

At the critical value x_c of x, these last two equations have a multiple solution at $s = 0$. Hence

$$-\frac{1 + 6x_c^2 + x_c^4}{2 + 2x_c^2} = -1,$$

i.e.

$$x_c = (\sqrt{5} - 2)^{1/2} = .4858 .$$

4. GENERAL FORM OF THE QUASI-CHEMICAL METHOD

Let us now take a group of any size and form. Let it have a α-sites b β-sites and γ pairs of nearest neighbours. The procedures to obtain an approximate expression for the free energy of the crystal follow exactly the same line as in the special case considered in the last section. Equations (13), (14) and (18) are essentially unchanged:

$$\overline{[q_1, q_2, \ldots]} = \xi \mu_1^{q_1} \mu_2^{q_2} \cdots e^{-\chi/kT}, \tag{31}$$

$$\phi = \sum_q \xi \mu_1^{q_1} \mu_2^{q_2} \cdots e^{-\chi/kT}, \quad \xi \frac{\partial \phi}{\partial \xi} = N, \quad \mu_i \frac{\partial \phi}{\partial \mu_i} = Nw, \tag{32}$$

and

$$(\mu_i)_{T=\infty} = \frac{w}{r}. \tag{33}$$

But (12) should be corrected by a factor $\frac{\gamma}{z}$ to account for the duplications in calculating E from the sum of the energies of all the groups in the crystal:

$$\overline{E} = \frac{z}{\gamma} \sum_q \overline{[q_1, q_2, \ldots]} \chi(q_1, q_2, \ldots) = \frac{z}{\gamma} kT^2 \frac{\partial \phi}{\partial T}. \tag{34}$$

Hence (22) becomes[*]

$$F(r, T) = -kT \left[\log g(r) + \frac{z}{\gamma} \Psi(r, T) - \frac{z}{\gamma} \Psi(r, \infty) \right], \tag{35}$$

or more explicitly, by (3), (17), and (33):

$$F(r, T) = -\frac{zNkT}{\gamma} \left[\log N + \left(a + b - \frac{2\gamma}{z} \right)(r \log r + w \log w) - \log \xi - w \sum_i \log \mu_i \right]. \tag{36}$$

---

[*]Care must be taken when the theory is extended to the case when the atomic ratio is not 1:1. The function Ψ in (35) must then be replaced by $\frac{1}{2}\Psi + \frac{1}{2}\Psi'$ where Ψ' is the function Ψ for the case when the group of interest has the same form as the original one but with α and β sites interchanged.

The derivative is

$$\frac{\partial}{\partial r}F(r,T) = -\frac{zNkT}{\gamma}\log\left[\left(\prod_i \mu_i\right)\left(\frac{r}{w}\right)^{a+b-\frac{2\gamma}{z}}\right], \tag{37}$$

so that the condition of equilibrium is

$$\prod_i \mu_i = \left(\frac{\overline{w}}{\overline{r}}\right)^{a+b-\frac{2\gamma}{z}}. \tag{38}$$

In actual calculations the following points may prove helpful:

(i) The free energy is changed by a constant if V_{AA} and V_{BB} are both replaced by $1/2(V_{AA}+V_{BB}) - V_{AB}$, and V_{AB} by 0. To prove this let z_i be the number of sites in the group neighbouring to the site i. Let χ be changed into χ' by the replacement. It is evident that

$$\chi' - \chi = -\gamma V_{AB} + \frac{V_{AA}-V_{BB}}{2}\text{ (no. of }B\text{-}B\text{ pairs}-\text{no. of }A\text{-}A\text{ pairs)},$$

and that

$$\sum_{\alpha\text{-sites}} q_i z_i - \sum_{\beta\text{-sites}} q_i z_i = \text{no. of }B\text{-}B\text{ pairs}-\text{no. of }A\text{-}A\text{ pairs}.$$

Hence

$$\overline{[q_1,q_2,\ldots]} = \xi\mu_1^{q_1}\mu_2^{q_2}\cdots e^{-\chi/kT} = \xi'\mu_1'^{q_1}\mu_2'^{q_2}\cdots e^{-\chi'/kT},$$

if we put

$$\xi' = \xi e^{-\gamma V_{AB}/kT}, \quad \mu_i' = \mu_i e^{\pm z_i(V_{AA}-V_{BB})/2kT},$$

where the + sign or the − sign is to be taken according as the site i is an α or a β site. We can now calculate the new free energy and verify the above statement.

(ii) Sites that are symmetrically situated in the group have equal μ's irrespective of their nature if $V_{AA} = V_{BB}$, $V_{AB} = 0$. This has already been shown in the last section. Since the most troublesome part of the calculations is the elimination of the parameters, much might be gained by choosing a group with a large number of sites symmetrically situated.

(iii) The free energy is a function of s^2, so that (38) is always satisfied at $\overline{w} = \overline{r} = 1/2$ (i.e. long distance order = 0). The proof is simple when we have already made $V_{AA} = V_{BB}$, $V_{AB} = 0$, so that an interchange of A and B atoms does not alter the energy. Thus

$$\chi(q_1,q_2,\ldots) = \chi(1-q_1, 1-q_2,\cdots).$$

Putting

$$\xi' = \xi\mu_1\mu_2\cdots$$

and
$$\mu'_i = \frac{1}{\mu_i}, \tag{39}$$
we get
$$\xi\mu_1{}^{q_1}\mu_2{}^{q_2}\cdots e^{-\chi/kT} = \xi'\mu_1'{}^{1-q_1}\mu_2'{}^{1-q_2}\cdots e^{-\chi/kT}.$$

Thus if (32) is satisfied
$$\sum_q (1-q_i)\xi'\mu_1'{}^{1-q_1}\mu_2'{}^{1-q_2}\cdots e^{-\chi/kT} = \sum_q \xi\mu_1{}^{q_1}\mu_2{}^{q_2}\cdots e^{-\chi/kT} - \sum_q q_i\xi\mu_1{}^{q_1}\mu_2{}^{q_2}\cdots e^{-\chi/kT}$$
$$= Nr,$$

i.e. ξ', μ'_1, μ'_2, ...would be the solution of (32) with r substituted for w. Hence by (32) and (17)
$$\Psi(1-r,T) = N - N\log\xi' - \sum_i Nr\log\mu'_i = \Psi(r,T)$$

showing that
$$F(1-r,T) = F(r,T). \tag{40}$$

(iv) <u>The parameter for a corner site is always given by</u>
$$\epsilon = \frac{1}{1+s}(\sqrt{x^2s^2 + (1-s^2)} - sx) \tag{41}$$

<u>irrespective of the size of the group, if $V_{AA} = V_{BB}$, $V_{AB} = 0$.</u> By a corner site we mean a site that has only one nearest neighbour in the group. Let ϵ be the selective variable (parameter) of a corner site, and μ_1 that of its only neighbour in the group. If the corner site is dropped, a new group is obtained. We distinguish all quantities referring to this new group by a prime, and obtain at once
$$\xi'\frac{\partial\phi'}{\partial\xi'} = N, \quad \mu'_i\frac{\partial\phi'}{\partial\mu'_i} = Nw, \quad i = 1,2,\cdots. \tag{42}$$

The sites of the primed group are numbered in the same way as in the unprimed group. Introducing the variable χ defined in (9) we may write
$$\phi = \sum_{p,q} \xi\epsilon^p\mu_1^{q_1}\mu_2^{q_2}\cdots e^{-\chi/kT}$$
$$= \sum_{q_2,\ldots} \xi(1+\epsilon x)\mu_2^{q_2}\mu_3^{q_3}\cdots e^{-\chi'/kT} + \mu_1\sum_{q_2,\ldots}\xi(\epsilon+x)\mu_2^{q_2}\cdots e^{-\chi'/kT}. \tag{43}$$

Let these two terms be denoted by ϕ_0 and ϕ_1 respectively. Since
$$\phi = N, \quad \mu_1\frac{\partial\phi}{\partial\mu_1} = Nw,$$
we have
$$\phi_0 = Nr, \quad \phi_1 = Nw. \tag{44}$$

Now
$$\epsilon\frac{\partial\phi_0}{\partial\epsilon} = \frac{\epsilon x}{1+\epsilon x}\phi_0, \quad \epsilon\frac{\partial\phi_1}{\partial\epsilon} = \frac{\epsilon}{\epsilon+x}\phi_1.$$

Hence $\epsilon\frac{\partial\phi}{\partial\epsilon} = Nw$ leads to
$$\frac{\epsilon x}{1+\epsilon x}Nr + \frac{\epsilon}{\epsilon+x}Nw = Nw, \tag{45}$$

or
$$\frac{w}{r} = \frac{\epsilon(\epsilon+x)}{1+\epsilon x}, \tag{46}$$

the solution of which is (41). Thus the two parameters μ and ν in the approximation discussed in section 2 are equal to ϵ.

(v) The "contribution" to the free energy from a corner atom is such that, in the notations of (iv).
$$F(r,T) = \frac{\gamma-1}{\gamma}F'(r,T) + \frac{1}{\gamma}F_0(r,T), \tag{47}$$

where $F_0(r,T)$ is the free energy when $\gamma = 1$, i.e. the free energy in the approximation discussed in section 2. This is proved as follows.

If we put
$$\xi = \xi''\frac{1}{1+\epsilon x}, \quad \mu_1 = \mu_1''\frac{1+\epsilon x}{\epsilon+x}, \quad \mu_i = \mu_i'', \ i \geq 2, \tag{48}$$

it is evident from (43) that ϕ would become a function of $\xi'', \mu_1'', \mu_2'', \cdots$ satisfying the relations
$$\xi''\frac{\partial\phi}{\partial\xi''}\left(=\xi\frac{\partial\phi}{\partial\xi}\right) = N, \quad \mu_i''\frac{\partial\phi}{\partial\mu_i''}\left(=\mu_i\frac{\partial\phi}{\partial\mu_i}\right) = Nw, \ i = 1, 2, \cdots. \tag{49}$$

It is also evident that ϕ is the same function of $\xi'', \mu_1'', \mu_2'', \cdots$ as ϕ' is of $\xi', \mu_1', \mu_2', \cdots$. Now (42) has only one[7] set of solution in ξ' and μ_i'. Hence from (49) we infer that $\xi' = \xi''$, $\mu_i' = \mu_i''$. Thus
$$\xi = \xi'\frac{1}{1+\epsilon x}, \quad \mu_1 = \mu_1'\frac{1+\epsilon x}{\epsilon+x}, \quad \mu_i = \mu_i', \ i \geq 2. \tag{50}$$

(41) and (50) give the parameters μ_i in terms of μ_i'. Inserting them into (36) we obtain

$$F(r,T) = -NkT\frac{z}{\gamma}\left[\log N + \left(a+b-\frac{2\gamma}{z}\right)(r\log r + w\log w) - \log\xi' - w\sum_i\log\mu_i' + r\log(1+\epsilon x) + w\log(\epsilon+x) - w\log\epsilon\right]$$
$$= \frac{\gamma-1}{\gamma}F'(r,T) - NkT\frac{z}{\gamma}\left[\left(1-\frac{2}{z}\right)(r\log r + w\log w) + r\log(1+\epsilon x) + w\log\frac{\epsilon+x}{\epsilon}\right]. \tag{51}$$

If the original (unprimed) group is a pair of nearest neighbours, we have $\gamma = 1$, and (51) reduces to the expression for the free energy in the approximation discussed in section 2:

$$F_0(r,T) = -zNkT\left[\left(1-\frac{2}{z}\right)(r\log r + w\log w) + r\log(1+\epsilon x) + w\log\frac{\epsilon + x}{\epsilon}\right]. \quad (52)$$

Inserting this back into (51) we get (47).

5. COMPARISON WITH BETHE'S METHOD

The so-called local grand partition function[1,3,8] in Bethe's approximations with long-distance order is identical in form with our function ϕ when all the "interior sites" in the group have the same parameter μ. For the case of equal concentrations for the two kinds of atoms, which is the case so far considered, this parameter has been put equal to unity by Bethe. Since the different terms of the local grand partition function stand for the probabilities of occurrence of the corresponding local groups in the crystal, it is clear that Bethe's method with long-distance order is essentially equivalent to our method plus the assumption that the free energy (35) has a minimum when

$$(\mu)_{\text{interior sites}} = 1. \quad (53)$$

But as we have shown that (38) gives the condition of a minimum of the free energy, the complete[*] identification of Bethe's and the quasi-chemical methods in any approximation reduces to the mathematical proof of the equivalence of (38) and (53). While this presents no difficulty at all for Bethe's first approximation (section 7), a general proof is by no means easy. We can only satisfy ourselves with the assertion that the two methods are equivalent for large groups, i.e. groups for which

$$a + b - \frac{2\gamma}{z} \ll \gamma.$$

This follows from the fact that if (53) is true

$$\left[\left(\prod \mu_i\right)\left(\frac{r}{w}\right)^{a+b-\frac{2\gamma}{z}}\right]^{\frac{1}{\gamma}} \cong \left[\prod (\mu)_{\text{interior sites}}\right]^{\frac{1}{\gamma}} = 1,$$

so that by (37)

$$\frac{\partial}{\partial r}F(r,T) = 0.$$

---

[8]Easthope, Proc. Camb. Phil. Soc. <u>33</u>, 502 (1937).
[*]"complete" as far as the probabilities of occurrence of the local configurations are concerned. The energy calculations are different in the two methods.

To see how the equilibrium free energy $F(\bar{r}, T)$ varies with T in high order approximations, we substitute (38) into (36) and make use of (32):

$$-\frac{F}{zNkT} = \frac{1}{\gamma} \log\left(\sum_q \mu_1^{q_1} \mu_2^{q_2} \cdots e^{-\chi/kT}\right) + \frac{1}{\gamma}\left(a + b - \frac{2\gamma}{z}\right) \log \bar{r}.$$

The last term is very small for large groups, so that by (53)

$$-\frac{F}{zNkT} = \frac{1}{\gamma} \log\left(\sum e^{-\chi/kT}\right).$$

6. THE NON-CINTERFERENCE OF LOCAL CONFIGURATIONS

Let us return to the fundamental assumption of the quasi-chemical method, i.e. to (31) which gives the average numbers of the different local configurations (so far called groups) in the crystal. This equation expresses the exact distribution law of an assembly of molecules (cf. the example in section 2) which has an energy γ/z times as large as the crystal. Distinguishing all quantities referring to the assembly of molecules by a subscript m, we get

$$F(r, T) + kT \log g(r) = \frac{z}{\gamma}[F_m(r, T) + kT \log g_m(r)],$$

which is obtained from (4). But if H is the number of arrangements in the crystal lattice having the given values of $[q_1, q_2, \ldots]$,

$$F(r, T) = -kT \log \overline{H} + \overline{E}. \tag{54}$$

Thus

$$\log \frac{\overline{H}}{g(r)} = \frac{z}{\gamma} \log \frac{\overline{H}_m}{g_m(r)}.$$

But[*]

$$H_m = \frac{N!}{\prod_q [q_1, q_2, \ldots]!}, \tag{55}$$

hence dropping the bar we get

$$H = h(r)\left\{\frac{N!}{\prod_q [q_1, q_2, \ldots]!}\right\}^{z/\gamma}, \tag{56}$$

where

$$h(r) = \frac{g(r)}{\{g_m(r)\}^{z/\gamma}}. \tag{57}$$

---
[*]Fowler, Statistical Mechanics, second edition, sections 2.6 and 5.11.

Equation (56) has been referred to in Fowler and Guggenheim's paper[1] as the mathematical expression of the "hypothesis of the non-interference of local configurations", because when $\gamma/z = 1$, the number of arrangements in the crystal consistent with the distribution law $[q_1, q_2, \ldots]$ for the groups of sites is, except for the factor $h(r)$, equal to

$$H_m = \frac{N!}{\prod_q [q_1, q_2, \ldots]!},$$

which is the number of arrangements in the crystal for the given values of $[q_1, q_2, \ldots]$ if the N groups in the crystal are <u>imagined</u> to be <u>separated</u> and are filled <u>independently</u> with atoms. The term "non-interference" comes from the fact that actually the N groups are <u>not separated</u> but are <u>interlocked</u> and <u>cannot be filled independently</u> with atoms, i.e. they "interfere" with each other.

To find the value of $g_m(r)$ we notice that by definition $g_m = \sum H_m$. But $\sum H_m$ is the number of arrangements in the N separated groups considered above if they are to be so filled with atoms that Nw of them have wrong atoms on the sites $i, i = 1, 2, \ldots$. Among the N sites i of the N groups $\frac{N!}{(Nr)!(Nw)!}$ different arrangements are possible. Hence*

$$g_m = \sum H_m = \sum \frac{N!}{\prod_q [q_1, q_2, \ldots]!} = \left[\frac{N!}{(Nr)!(Nw)!} \right]^{a+b}. \tag{58}$$

Thus

$$h(r) = \left[\frac{N!}{(Nr)!(Nw)!} \right]^{2-(a+b)\frac{z}{\gamma}}. \tag{59}$$

The free energy of the crystal may be obtained from (54), (56) and (59):

$$F(r,T) = \overline{E} - \frac{zNkT}{\gamma} \left\{ \left(a + b - \frac{2\gamma}{z} \right) (r \log r + w \log w) + \log N - \frac{1}{N} \sum_q \overline{[q_1, q_2, \ldots]} \log \overline{[q_1, q_2, \ldots]} \right\}, \tag{60}$$

which has been obtained above by integration.

---

*It might be mentioned in passing that for the special case considered in section 2. (58) gives directly the value of the sum $\sum_x$ in equation (8.5) of Fowler and Guggenheim's paper if their r and q are equal. The generalization to the case r/q is however easy. The result is

$$\sum_x \frac{[zN]!}{[zN(r-x)]![zNx]![zN(1-r-q+x)]![zN(q-x)]!} = \frac{[zN]!}{[zNr]![zN(1-r)]!} \frac{[zN]!}{[zNq]![zN(1-q)]!} \tag{60}$$

which is exact. The value of $\log \sum_x$ given by (60) reduces to the approximate expression that Fowler and Guggenheim obtained by identifying $\sum_x$ with its maximum term when N is large.

7. SPECIAL CONSIDERATIONS CONCERNING BETHE'S FIRST AND SECOND APPROXIMATIONS

(i) <u>First Approximation</u>. If an α-site together with its z nearest neighbours are taken as our group of interest, all the sites except the central one are corner sites. Hence their selective variables are all equal to the value of ϵ given in (41). By successive applications of (47) we see that the free energy is exactly $F_0(r,T)$, a fact which has already been pointed out by Fowler and Guggenheim[1]. The selective variable of the central site is given by successive applications of (50)

$$\lambda = \frac{w}{r}\left(\frac{1+\epsilon x}{\epsilon + x}\right)^z. \tag{61}$$

The factor w/r is the selective variable for the central site when it alone forms the group. The equilibrium condition (38) becomes

$$\lambda \epsilon^z = \left(\frac{\overline{w}}{\overline{r}}\right)^{z-1}.$$

But by (61) and (45),

$$\lambda = \left(\frac{w}{r}\right)^{1-z} \epsilon^z.$$

Hence at equilibrium

$$\lambda = 1. \tag{62}$$

Thus the approximation is completely equivalent to Bethe's first approximation, as already mentioned in section 5.

(ii) <u>Second Approximation</u>. Now consider the group of sites occurring in Bethe's second approximation[3]. According to section 4, (iv), the selective variables for the corner sites in the second shell are all equal to ϵ, which is given by (41). But in Bethe's original calculations, the selective variables for the corner sites and the medium sites are made equal, and are found to be different from ϵ. Thus if we use his original method, equation (32) can not be satisfied. (In other words, the probabilities of occurrence or wrong atoms in the corner and the medium sites would be unequal.)

For simplicity we shall drop the corner sites and take as our group of interest the central site, the first shell sites and the medium sites; with selective variables μ, ν and λ respectively. (The contribution by the corner sites can be included in the free energy by simple addition as shown in section 4 (v).) With the notations n, and g_{nm} of Bethe[3] we find

$$\phi = \xi \sum_n (x^n + \mu x^{z-n}) P_n(x, \nu, \lambda), \tag{63}$$

Investigations in the Statistical Theory of Superlattices

where

$$P_n(x, \nu, \lambda) = \nu^n \sum_m g_{nm}[(1+\lambda)x]^m (x^2+\lambda)^{(\frac{z}{2}-1)n-\frac{m}{2}}(1+\lambda x^2)^{(\frac{z}{2}-1)(z-n)-\frac{m}{2}}.$$

After eliminating ξ and μ, (32) becomes

$$zw = \frac{2\sum_{l,n} x^{l-n}\left(rP_n\lambda\frac{\partial}{\partial\lambda}P_l + wP_l\lambda\frac{\partial}{\partial\lambda}P_n\right)}{(z-2)\left(\sum_n x^n P_n\right)\left(\sum_n x^{-n} P_n\right)} = \frac{\sum_{l,n} x^{l-n}\left(rP_n\nu\frac{\partial}{\partial\nu}P_l + wP_l\nu\frac{\partial}{\partial\nu}P_n\right)}{\left(\sum_n x^n P_n\right)\left(\sum_n x^{-n} P_n\right)}. \quad (64)$$

The free energy is obtained from (36):

$$F(r,T) = -\frac{NkT}{z-1}\left[\frac{1}{2}(z^2-4z+4)(r\log r + w\log w) + r\log\sum_n x^n P_n + w\log\sum_n x^{z-n} P_n - zw\log\nu - wz\left(\frac{z}{2}-1\right)\log\lambda\right]. \quad (65)$$

8. APPLICATION TO THE CRYSTAL Cu_3Au

For the face-centred crystal Cu_3Au, we may of course follow Peierls[9] and take as our group a central site together with its twelve first shell neighbours. The free energy expression would then contain seven selective variables*, four of which can be eliminated. The resultant expression is very cumbersome and numerical calculations would be laborious. We therefore make a simpler approximation: the group is taken to be four nearest neighbours forming a tetrahedron. A little geometrical consideration assures us that all such tetrahedrons contain an α-site (for gold atoms) and three β-sites (for copper atoms), an interesting conclusion showing that the tetrahedron might be

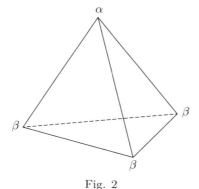

Fig. 2

---

[9] Peierls, Proc. Rey. Soc. London A154, 207(1936).

*For α-centred groups, three, and for β-centred, four, parameters are necessary. Both these two kinds of groups must be considered because otherwise the energy of the crystal cannot be easily obtained from the energy of the groups in the crystal.

regarded as a sort of "molecular" structure in a face-centred lattice with atomic ratio 1:3. Our approximation may thus be reasonably expected to reveal the more important features of order–disorder transformation in such alloys.

Let μ and ν be the parameters (for wrong atoms) of the β-sites and the α-sites respectively. Let there be altogether $4N$ atoms. It is easy to see that there are $8N$ groups in the crystal. When Nw atoms on the α-sites are wrong, the equations determining the parameters are

$$8N = \phi = \xi[x^3 + 3x^2\mu + 3x^3\mu^2 + x^6\mu^3 + \nu(x^6 + 3x^3\mu + 3x^2\mu^2 + x^3\mu^3)], \tag{66a}$$

$$8Nw = \nu\frac{\partial \phi}{\partial \nu} = \xi\nu(x^6 + 3x^3\mu + 3x^2\mu^2 + x^3\mu^3), \tag{66b}$$

and

$$8N\left(\frac{w}{3}\right) + 8N\left(\frac{w}{3}\right) + 8N\left(\frac{w}{3}\right) = \mu\frac{\partial \phi}{\partial \mu} = 3x^2\mu\xi[1 + 2x\mu + x^4\mu^2 + \nu(x + 2\mu + x\mu^2)], \tag{66c}$$

where x is defined by (9). The energy of the crystal is (cf. (34)),

$$\overline{E} = \frac{1}{2}kT^2\frac{\partial \phi}{\partial T} + \text{constant}; \tag{67}$$

so that the free energy becomes (cf. (35))

$$F(w,T) = -kT\left[\log g(w) + \frac{1}{2}(\phi - 8N\log\xi - 8Nw\log\nu - 8Nw\log\mu)_{T=\infty}^T\right].$$

But

$$\log g(w) = -N\left\{(1-w)\log(1-w) + w\log w + w\log\frac{w}{3} + (3-w)\log[(3-w)/3]\right\},$$

and at $T = \infty$,

$$\nu = \frac{w}{1-w}, \quad \mu = \frac{w}{3-w}, \quad \xi = 8N(1-w)\left(1 - \frac{w}{3}\right)^3.$$

Hence

$$-\frac{F(w,T)}{NkT} = -9\log 3 + 4\log 8N + 6w\log w + 3(1-w)\log(1-w) +$$
$$3(3-w)\log(3-w) - 4\log\xi - 4w\log\mu - 4w\log\nu. \tag{68}$$

Since ξ and ν can be very easily solved from (66), numerical calculations are quite simple. The equilibrium value of w is given by (cf. (37) and (38))

$$0 = -3\log\frac{(1-\overline{w})(3-\overline{w})}{\overline{w}^2} + 4\log\mu\nu. \tag{69}$$

This is always satisfied at $\overline{w} = \frac{3}{4}$[*]. Actual calculation shows that the absolute minimum

---

[*]This is not evident from (69) directly. But if we divide the whole crystal into four sublattices which are all simple cubic and introduce a w for each sublattice so that Nw is the number of A atoms on the i-th sublattice ($i = 1, 2, 3, 4$), it is obvious that the free energy is symmetrical in the w's. From this we infer that (69) is satisfied at $\overline{w} = \frac{3}{4}$.

of the free energy is or is not at $\bar{w} = \frac{3}{4}$ according as $x \geq .2965$ or $x < .2965$. The value of the free energy is plotted in Fig.3. From the form of the graph it is seen that the crystal has a critical temperature at which the long-distance order and (hence) the energy are discontinuous. The critical temperature T_C and the latent heat Q are found to be

$$T_C = .8228 \frac{1}{k} \left[\frac{1}{2}(V_{AA} + V_{BB}) - V_{AB} \right], \quad Q = .8824N \left[\frac{1}{2}(V_{AA} + V_{BB}) - V_{AB} \right].$$

In terms of the total energy change from $T = 0$ to $T = \infty$:

$$E_0 = 3N \left[\frac{1}{2}(V_{AA} + V_{BB}) - V_{AB} \right],$$

these quantities become

$$T_C = 1.097 E_0/R, \quad (T_C = 2.19 E_0/R \text{ in Bragg-Williams' approximation and}$$
$$T_C \cong 1.3 E_0/R \text{ in Peierls' approximation.})$$
$$Q = .2941 E_0, \quad (Q = .218 E_0 \text{ in Bragg-Williams' approximation and}$$
$$Q \cong .36 E_0 \text{ in Peierls' approximation.})$$

where R stands for $4Nk$.

It will be noticed that due to the lack of a free energy expression Peierls[9] did not give the exact values of these quantities.

In conclusion, the author wishes to express his thanks to Prof. J. S. Wang for valuable criticism and advice.

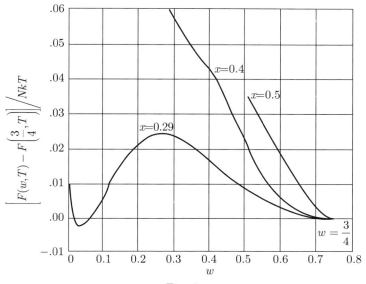

Fig. 3

National Tsing Hua University

INVESTIGATIONS IN THE STATISTICAL THEORY OF SUPERLATTICES

A Dissertation
Submitted to
The Faculty of the Graduate School of Science
in Candidacy for
The Degree of Master of Science

By   C. N. Yang   (楊振寧)

Kunming, China
June, 1944

CONTENTS

I. The Variation of the Interaction Energy with Change of Lattice Constants and Change of the Degree of Order

II. A Generalization of the Quasi-Chemical Method in the Statistical Theory of Superlattices

THE VARIATION OF THE INTERACTION ENERGY WITH CHANGE OF LATTICE CONSTANTS AND CHANGE OF THE DEGREE OF ORDER

By C. N. Yang
National Tsing Hua University,
Kunming, China

ABSTRACT

The change of the lattice constants due to the order-disordering process in a superlattice is investigated by using the condition of minimum free energy in Bethe's theory. It is found that the interaction energy depends on the degree of order when the external pressure is kept constant. The specific heat at constant pressure given by the theory is compared with experiment. Another cause of the variation of interaction energy is the change of atomic arrangements. This is also investigated from the view point of Wang's formulation of the free energy in Bethe's approximations.

1. INTRODUCTION

The binary alloy CuAu is face-centred cubic when disordered and tetragonal when ordered. This change of lattice form can be studied thermodynamically if we know the energy and the entropy of the crystal. Some calculations along this line has already been made by Wilson[1] who used Bethe's method to find the energy but Bragg-Williams' method to find the entropy of the crystal. It will be shown in the present paper that Bethe's method can be carried through in the calculations, making it self-consistent. The results are comparable with Gorsky's measurements[2].

The change of lattice constants evidently affects the interaction energy between the atoms, and must consequently produce a change in the configurational energy and the specific heat of the crystal. We shall see that the effect is in the right direction to bring the theory into closer agreement with experiment, because it tends to make the energy increase more rapidly near the critical temperature. An actual calculation of the specific heat at variable lattice constant but constant external pressure for β-brass is given in section 3.

Now the interaction energy can also be influenced by the change of the atomic arrangements. Mott[3] has shown from a study of the electronic distribution in superlattices that the interaction energy decreases as the degree of order decreases. The actual relation between the two is naturally very complicated. A linear dependence (of the average interaction energy upon the degree of order) has been assumed by Lin[4] in attempting to explain the occurrence of the maximum critical temperature of a face-centred alloy at the concentration ratio 1:3. In order to justify the assumption we shall view the problem from a new angle by the introduc-

---
[1] Wilson, Proc. Camb. Phil. Soc. **34**, 81 (1938).
[2] Gorsky, Zeit. f. Phys. **50**, 64 (1928).
[3] Mott, Proc. Phys. Soc. **49**, 258 (1937).
[4] Lin, Chinese J. Phys. **3**, 182 (1939).

-2-

tion of the free energy in Bethe's approximation[5]. In this way it is found that the interaction energy as a function of the degree of order must satisfy certain equations obtained from a set of conditions of consistency. This same set of conditions of consistency makes also possible the calculation of the energy of the crystal without appealing to Bragg-Williams' theory as Lin did.

2. THE VARIATION OF LATTICE CONSTANTS

We shall form the partition function at constant lattice constants l_1 and l_2, and then obtain their equilibrium values from the equations determining the generalized reactions. Let $\tfrac{1}{2} z N m$ be the number of A-B neighbours in the crystal. If $g(m)$ is the number of arrangements of the atoms for the given value of m, and $W(l_1, l_2, m)$ the configurational energy of the crystal, the configurational partition function is

$$f(m,T,l_1,l_2) = g(m)\exp(-W/kT).$$

The equilibrium value $\bar{m}$ of m is determined from the condition of a maximum of f:
$$\frac{\partial}{\partial \bar{m}} \log f(\bar{m},T,l_1,l_2) = 0.$$

The generalized reactions are given by

$$L_i = kT \frac{d}{dl_i} \log f(\bar{m},T,l_1,l_2) = kT \frac{d\bar{m}}{dl_i}\frac{\partial}{\partial \bar{m}}\log f + kT\frac{\partial}{\partial l_i}\log f = kT\frac{\partial}{\partial l_i}\log f$$
$$= -\frac{\partial}{\partial l_i} W(l_1,l_2,\bar{m}) \ . \tag{1}$$

To study the change of lattice form in CuAu we divide the face-centred lattice into four simple cubic sublattices 1,2,3,4*. Let the shortest distance between the sites of 1 and 2, or 3 and 4 be l_1, that between the sites of 1 and 3, 1 and 4, 2 and 3 or 2 and 4 be l_2, so that the former is the distance between neighbouring Au-Au or Cu-Cu atoms and the latter that between neighbouring Au-Cu atoms when the crystal is perfectly ordered. The interaction energies V_{AA}, V_{AB} and V_{BB} are functions of l_1 and l_2.

---
*Cf. Fig. 27 in Rev. Mod. Phys. 10, 1(1938).
[5] Wang, "Free Energy in the Statistical Theory of Order-Disorder Transformations", Science Report of National Tsing Hua University, series A, 30-th anniversary Memorial Number (1941), printed but failed to appear.

-3-

If the number of sites of each sublattice is $\frac{1}{2}N$, the number of pairs of sites between sublattices 1 and 2 must be $4(\frac{1}{2})N = 2N$. Denote by m_{12} the fraction of A-B pairs among these. Then the number of

A-A pairs is $\quad \frac{1}{2}[4(\frac{N}{2}\theta_1 + \frac{N}{2}\theta_2) - 2Nm_{12}] = N[\theta_1 + \theta_2 - m_{12}]$,

B-B pairs is $\quad \frac{1}{2}[4(\frac{N}{2}\{1-\theta_1\} + \frac{N}{2}\{1-\theta_2\}) - 2Nm_{12}] = N[2 - \theta_1 - \theta_2 - m_{12}]$,

where θ_i is the fraction of sites of sublattice i occupied by A atoms. Thus the energy of interaction between the atoms on sublattices 1 and 2 is

$$N[(\theta_1 + \theta_2 - m_{12})V_{AA}(l_1) + 2m_{12}V_{AB}(l_1) + (2 - \theta_1 - \theta_2 - m_{12})V_{BB}(l_1)] .$$

Writing $\quad c = \frac{1}{4}(\theta_1 + \theta_2 + \theta_3 + \theta_4) \quad$ and $\quad V = \frac{1}{2}(V_{AA} + V_{BB}) - V_{AB}$

we get the energy of the whole crystal

$$W = N[4cV_{AA}(l_1) + 4(1-c)V_{BB}(l_1) - 2(m_{12} + m_{34})V(l_1)]$$
$$+ N[8cV_{AA}(l_2) + 8(1-c)V_{BB}(l_2) - 2(m_{13} + m_{14} + m_{23} + m_{24})V(l_2)] . \quad (2)$$

With this value for W, (1) becomes

$$L_1 = -N[4cV_{AA}'(l_1) + 4(1-c)V_{BB}'(l_1) - 2(\bar{m}_{12} + \bar{m}_{34})V'(l_1)] , \quad (3)$$

and $\quad L_2 = -N[8cV_{AA}'(l_2) + 8(1-c)V_{BB}'(l_2) - 2(\bar{m}_{13} + \bar{m}_{14} + \bar{m}_{23} + \bar{m}_{24})V'(l_2)] . \quad (4)$

To solve for l_1 and l_2 as functions of T we must first know the $\bar{m}$'s, which are usually very complicated. Wilson[1] discussed the values of l_1 and l_2 only in the cases when the alloy is disordered and when the order is nearly perfect. We shall also confine our attention to these cases.

(i) <u>Disordered</u>. In this case there is no difference between the four sublattices so that all the $\bar{m}_{ij}$'s are equal to $\bar{m}$. (3) and (4) reduce to

$$L_1 = -N[4cV_{AA}'(l_1) + 4(1-c)V_{BB}'(l_1) - 4\bar{m}V'(l_1)] ,$$
$$L_2 = -N[8cV_{AA}'(l_2) + 8(1-c)V_{BB}'(l_2) - 8\bar{m}V'(l_2)] . \quad (5)$$

If $L_1 = L_2 = 0$, this shows that $l_1 = l_2$, so that the crystal is cubic.

(ii) <u>Order nearly perfect</u>. When $c = \frac{1}{2}$, and the order is nearly perfect,

$$\theta_1 = \theta_2 \cong 1, \quad \theta_3 = \theta_4 \cong 0, \quad \theta_1 + \theta_3 = 1, \quad \theta_1 - \theta_3 = s.$$

There are only a few B atoms on sublattices 1 and 2. Hence approximately

$$\bar{m}_{12} = (1-\theta_1) + (1-\theta_2) = 2\theta_3 = 1-s.$$

By the same reason we can obtain the number of A-A pairs of neighbours

between the sublattices 1 and 3:
$$N(\theta_1+\theta_3-\overline{m}_{13})=4(\tfrac{1}{2}N\theta_3).$$
Thus
$$\overline{m}_{13}=\theta_1-\theta_3=s.$$
We can now write down all the $\overline{m}$'s:
$$\overline{m}_{12}=\overline{m}_{34}=1-s, \quad \overline{m}_{13}=\overline{m}_{23}=\overline{m}_{14}=\overline{m}_{24}=s.$$
These equations are correct to the first order of $(1-s)$. Substituting them into (3) and (4) we obtain
$$L_1=-2N[V_{AA}'(l_1)+V_{BB}'(l_1)-2(1-s)V'(l_1)],$$
and
$$L_2=-4N[V_{AA}'(l_2)+V_{BB}'(l_2)-2sV'(l_2)].$$
These are exactly equations (27) in Wilson's paper, from which an expression of the degree of tetragonality in agreement with Gorsky's measurements[2] can be obtained.

3. THE EFFECT OF THE CHANGE OF LATTICE CONSTANTS ON THE INTERACTION ENERGY

In the alloy CuAu the gold atoms and the copper atoms are in contact when the order is perfect. Since the copper atom is somewhat smaller than the gold atoms, the size of the crystal must increase when gold atoms exchange their positions with copper atoms. Thus with increasing disorder the distance between the atoms increases and hence the interaction energies diminish. The disordering process is therefore effected with more ease near the critical temperature than it is at lower temperatures; and we expect the specific heat at constant pressure to possess a steeper and higher maximum at the critical temperature than the specific heat at constant volume.

Now we shall calculate in length the specific heat at constant pressure of the alloy β-brass, which forms the simplest type of superlattice that can be studied statistically. Bethe's method will be used.

The configurational energy of the crystal is, in Easthope's[6] notations:
$$W=-N_{AB}V+\tfrac{1}{2}Nz\{c(V_{AA}-V_{BB})+V_{BB}\}. \tag{6}$$

---
[6] Easthope, Proc. Camb. Phil. Soc. 33, 502(1937); 34, 68(1938).

Substitution of this expression into (1) gives

$$0 = -\bar{m}V'(1) + c[V_{AA}'(1) - V_{BB}'(1)] + [V_{BB}'(1)],$$

when the pressure is put equal to zero. Now the variation of V is not very large, so that to a sufficient approximation we may assume the linear relations

$$(V_{AA}'(1) - V_{BB}'(1))/V'(1) = -K_o + K_1 V, \quad (7)$$

and

$$V_{BB}'(1)/V'(1) = -J_o + J_1 V. \quad (8)$$

These three last equations give, after eliminating $V_{AA}'(1)$ and $V_{BB}'(1)$:

$$V = \frac{\bar{m} + (cK_o + J_o)}{cK_1 + J_1}. \quad (9)$$

We have already seen that V increases as $\bar{m}$ increases, hence $cK_1 + J_1$ must be positive. The other constant $cK_o + J_o$ must also be positive in order that V may be positive with only a relatively small variation.

Eisenschitz[7] has calculated the specific heat at constant pressure by Bragg-Williams' method. He assumed that the interaction energies depends on a parameter u in the following way:

$$\tfrac{1}{2}(V_{AA} + V_{BB}) = \phi[(1-a) + a(1-u)^2], \quad V_{AB} = \phi b u^2,$$

where $a = .225$, and $b = .203$ and u is of the order of unity. Comparing this with (7) and (8) we see that his assumption is equivalent (approximately) to ours if $\tfrac{1}{2}K_o + J_o = 1.22$ and $\tfrac{1}{2}K_1 + J_1 = .508 \times 10^{14}$ erg^{-1}. But with these values the specific heat at the critical temperature would be too large. In order to make $(C_p)_{T_c} = 5.1R$ as given by the measurements of Sykes and Wilkinson[3] we assume (cf. eq.(12) below)

$$\tfrac{1}{2}K_o + J_o = 1.79.$$

With this value for $\tfrac{1}{2}K_o + J_o$, the relative variation of V can be shown to be within 1.3%.

We can now start from (9) and the equations given by Easthope[6] for the determination of $\bar{m}$ as a function of the temperature and V to obtain the specific heat at constant pressure:

[7] Eisenschitz, Proc. Roy. Soc. 68, 546(1938).
[3] Sykes and Wilkinson, Inst. Metals J. 61, 223(1937).

-6-

$$C_p = \frac{dW}{dT} = \left(\frac{\partial W}{\partial \bar{m}}\right)_{\ell} \frac{d\bar{m}}{dx} \frac{dx}{dT} + \left(\frac{\partial W}{\partial \ell}\right)_{\bar{m}} \frac{d\ell}{dT} = \left(\frac{\partial W}{\partial \bar{m}}\right)_{\ell} \frac{d\bar{m}}{dx} \frac{dx}{dT} \quad . \tag{10}$$

But
$$\frac{dx}{dT} = (xV/kT^2)/(1 + \frac{x}{kT}\frac{dV}{d\bar{m}}\frac{d\bar{m}}{dx}) \quad . \tag{11}$$

Hence by (6)
$$C_p = \frac{\frac{1}{2}Nzkx(\log x)^2(-\frac{d\bar{m}}{dx})}{1 + \frac{x\log x}{\bar{m} + (cK_0 + J_0)}(-\frac{d\bar{m}}{dx})} \quad . \tag{12}$$

The value of this expression is calculated for the case $\sigma = \frac{1}{2}$, the constant $\frac{1}{2}K_0 + J_0$ being assumed to be 1.79 to make $(C_p)_{T_C} = 5.1R$. The result is plotted in the accompanying figure together with Bethe's curve for C_V[9] and Sykes and Wilkinson's experimental[8] data.

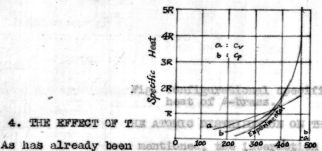

Fig. configurational specific heat of β-brass.

4. THE EFFECT OF THE ATOMIC ARRANGEMENT ON THE INTERACTION ENERGY

As has already been mentioned the interaction energy depends in some very complicated manner upon the degree of order. To study the effect of such a dependence Lin[4] has assumed a linear relationship:

$$V = V_0(1 + \alpha c + \beta m_{AA}) \tag{13}$$

between the interaction energy V and the fraction of A-A pairs of neighbours m_{AA}. In this section we shall study the general nature of the variation of V in the light of the theory of the free energy in Bethe's approximation given by Wang[5].

The fundamental equations in Wang's paper are (45), (46) and (39) with ξ_α and ξ_β given by (47), (48), (49) and (50). These equations are still assumed to be valid now V becomes a function of θ_α, θ_β and T. They may be put into the form:

$$\frac{d}{d\theta_\alpha}\log Q = -Nr_\alpha \log \xi_\alpha = \frac{\partial}{\partial \theta_\alpha}\log Q_0(\theta_\alpha, \theta_\beta, V/T) \quad , \tag{14}$$

---
[6] Nix and Shockley, Rev. Mod. Phys. 10, 1(1938).

-7-

$$\frac{d}{d\theta_\beta}\log Q = -Nr_\beta \log \frac{r_\beta}{l_\beta} = \frac{\partial}{\partial \theta_\beta}\log Q_0(\theta_\alpha,\theta_\beta,V/T) \quad , \tag{15}$$

$$E = kT^2 \frac{d}{dT}\log Q \quad , \tag{16}$$

where Q_0 is the partition function for the case when V is a constant, if we denote by $\frac{\partial}{\partial \theta_\alpha}$, $\frac{\partial}{\partial \theta_\beta}$, $\frac{\partial}{\partial T}$ and $\frac{\partial}{\partial V}$ differentiations when θ_α, θ_β, T and V are regarded as independent of each other, and by $\frac{d}{d\theta_\alpha}$ the operator $\frac{\partial}{\partial \theta_\alpha} + \frac{dV}{d\theta_\alpha}\frac{\partial}{\partial V}$. (14) and (15) mean that we have assumed with Lin that the equilibrium values of θ_α and θ_β are given by the same equations as in Bethe's approximation. (16) gives the energy of the crystal.

Consistency of (14) and (15) requires

$$\frac{d}{d\theta_\alpha}\left(\frac{\partial}{\partial \theta_\beta}\log Q_0\right) = \frac{d}{d\theta_\beta}\left(\frac{\partial}{\partial \theta_\alpha}\log Q_0\right) \quad ,$$

i.e.
$$\frac{dV}{d\theta_\alpha}\frac{\partial^2}{\partial \theta_\beta \partial V}\log Q_0 = \frac{dV}{d\theta_\beta}\frac{\partial^2}{\partial V \partial \theta_\alpha}\log Q_0 \quad . \tag{17}$$

Upon the hypothesis of nearest neighbour interaction the energy in Wang's paper becomes
$$E_0 = kT^2 \frac{\partial}{\partial T}\log Q_0 = zNVm_{AA} \quad .$$

But Q_0 depends on T and V through V/T, so that

$$\frac{\partial^2}{\partial \theta_\alpha \partial V}\log Q_0 = -\frac{T}{V}\frac{\partial^2}{\partial T \partial \theta_\alpha}\log Q_0 = -\frac{\partial E_0}{\partial \theta_\alpha}/kTV \quad .$$

Substituting this eq. and a similar one into (17) we obtain

$$\frac{\partial E_0}{\partial \theta_\beta}\frac{dV}{d\theta_\alpha} = \frac{\partial E_0}{\partial \theta_\alpha}\frac{dV}{d\theta_\beta} \quad , \tag{18}$$

which becomes, if $\frac{\partial E_0}{\partial V}\frac{dV}{d\theta_\alpha}\frac{dV}{d\theta_\beta}$ is added to both sides,

$$\frac{dE_0}{d\theta_\alpha}\frac{dV}{d\theta_\beta} = \frac{dE_0}{d\theta_\beta}\frac{dV}{d\theta_\alpha} \quad .$$

This shows that V and E_0 are connected by a relation independent of θ_α and θ_β, i.e.
$$V = V(E_0, T) \quad . \tag{19}$$

A consequence of this result is that the coefficient α in Lin's relation (13) must be zero. This makes, however, the maximum critical temperature for the AB type of superlattice to shift to a value of the concentration different from $\frac{1}{2}$, which contradicts experimental results. One way out of the difficulty is to make some other assumption regarding the dependence of V on the degree of order, such as

$$V = V_0\left[1 + \beta m_{AA} - \frac{\beta(z-1)}{2c(cz-1)}m_{AA}^2\right] \quad . \tag{20}$$

-8-

Let us now try to find E in the general case. From (14) and (16) we get

$$\frac{dE}{d\theta_\alpha} = kT^2 \frac{d}{dT}\left(\frac{\partial}{\partial \theta_\alpha}\log Q_o\right) = kT^2 \frac{d(\frac{V}{T})}{dT}\frac{\partial^2}{\partial \theta_\alpha \partial(\frac{V}{T})}\log Q_o = -\frac{T^2}{V}\frac{d(\frac{V}{T})}{dT}\frac{\partial E_o}{\partial \theta_\alpha}. \quad (21)$$

Similarly $\quad \frac{dE}{d\theta_\beta} \quad = -\frac{T^2}{V}\frac{d(\frac{V}{T})}{dT}\frac{\partial E_o}{\partial \theta_\beta}.$

Hence $\quad \frac{\partial E}{\partial \theta_\alpha}\frac{\partial E_o}{\partial \theta_\beta} = \frac{\partial E_o}{\partial \theta_\alpha}\frac{\partial E}{\partial \theta_\beta}.$

Just as (18) leads to (19), this last equation leads to

$$E = E(E_o, T).$$

Substituting this into (21) we obtain

$$\frac{\partial E}{\partial E_o}\left[\frac{\partial E_o}{\partial \theta_\alpha} + \frac{\partial E_o}{\partial V}\frac{\partial V}{\partial \theta_\alpha}\right] = \frac{\partial E_o}{\partial \theta_\alpha}\left(1 - \frac{T}{V}\frac{dV}{dT}\right).$$

But (19) gives $\quad \frac{dV}{d\theta_\alpha} = \frac{\frac{\partial E}{\partial \theta_\alpha}\frac{\partial V}{\partial E_o}}{1 - \frac{\partial V}{\partial E_o}\frac{\partial E_o}{\partial V}} \quad$ and $\quad \frac{dV}{dT} = \frac{\frac{\partial V}{\partial T} + \frac{\partial V}{\partial E_o}\frac{\partial E_o}{\partial T}}{1 - \frac{\partial V}{\partial E_o}\frac{\partial E_o}{\partial V}}.$

Hence $\quad \frac{\partial E}{\partial E_o} = \left(1 - \frac{\partial V}{\partial E_o}\frac{\partial E_o}{\partial V}\right)\left(1 - \frac{T}{V}\frac{dV}{dT}\right) = 1 - \frac{\partial V}{\partial E_o}\frac{\partial E_o}{\partial V} - \frac{T}{V}\frac{dV}{dT} - \frac{T}{V}\frac{\partial V}{\partial E_o}\frac{\partial E_o}{\partial T}.$

Now E_o/V is a function of $\theta_\alpha, \theta_\beta$ and V/T, so that

$$\frac{\partial E_o}{\partial V} = -\frac{T}{V}\frac{\partial E_o}{\partial T} + \frac{E_o}{V}.$$

Hence $\quad \frac{\partial E}{\partial E_o} = 1 - \frac{E_o}{V}\frac{\partial V}{\partial E_o} - \frac{T}{V}\frac{\partial V}{\partial T}. \quad (22)$

If V depends on m_{AA} only, and not on T, we have

$$\frac{\partial E}{\partial E_o} = 1 - \frac{E_o}{V}\frac{dV}{dE_o} = V\frac{d(E_o/V)}{dE_o}.$$

The boundary condition is given by the case when there is no A atoms, i.e. when $\theta_\alpha + \theta_\beta = 0$. In this case $m_{AA} = 0$, $E_o = E = 0$. Hence

$$E = \int_0^{E_o}\left(V\frac{d(E_o/V)}{dE_o}\right)dE_o = \int E \, d\cdot\log(E/V) = zN\int_0^{m_{AA}} V dm_{AA}. \quad (23)$$

If the assumption (20) is made, the energy of the crystal is

$$E = zNV_o\left[m_{AA} + \frac{\beta}{2}m_{AA}^2 - \frac{\beta(z-1)}{6c(cz-1)}m_{AA}^3\right].$$

When β is not large, this differs very little from Bethe's original expression in numerical value. The specific heat is

$$C_V = zNV\frac{dm_{AA}}{dT} = zNV\frac{dm_{AA}}{dx}\frac{\frac{xV}{kT^2}}{1 + \frac{x}{kT}\frac{dV}{dx}} = zNkx(\log x)^2\frac{dm_{AA}}{dx}\frac{1}{1 + \frac{x}{kT}\frac{dV}{dm_{AA}}\frac{dm_{AA}}{dx}}.$$

The author is very much indebted to Prof. J. S. Wang for suggesting this problem and for helpful discussions.

A GENERALIZATION OF THE QUASI-CHEMICAL METHOD IN THE STATISTICAL THEORY OF SUPERLATTICES

By C. N. Yang[*]
National Tsing Hua University,
Kunming, China

ABSTRACT

The quasi-chemical method introduced by Fowler and Guggenheim for the equilibrium distribution of pairs of sites in a superlattice is generalized. It is shown that by considering groups containing large numbers of sites the method may be used to obtain successive approximations of the free energy of the crystal. To analyze the fundamental assumption underlying the method more closely, the hypothesis of the non-interference of local configurations is discussed. The free energy of the crystal is obtained without integration as a closed algebraic expression with the aid of a Legendre transformation. Applications of the results are then made to different approximations for simple and body-centred cubic crystals and for the face-centred cubic crystal Cu_3Au. In each case the free energy is obtained and discussed.

---

[*]Research Fellow of the China Foundation for the Promotion of Education and Culture.

1. INTRODUCTION

It was shown by Fowler and Guggenheim[1] that the quasi-chemical method, originally devised for the theory of regular solutions, applies equally well to the theory of superlattices with long-distance order. The method is, as they have emphasized, definitely one stage further towards an exact theory than Bragg-Williams' method[2]. When compared with Bethe's[3] or Kirkwood's[4] method it also distinguishes itself in mathematical simplicity. But to be a method that may lead to a consistent scheme of successive approximations, it must be applicable to the n-th approximation in the theory of superlattices. This does not, however, seem possible in the original form of the method given by Fowler and Guggenheim. It is the purpose of the present paper to formulate the quasi-chemical method in a new way which is applicable to high order approximations in the theory of superlattices.

The free energy expression in Bethe's and in the quasi-chemical methods involves an integral. Its evaluation is very complicated and has been carried out[1,5] so far only in Bethe's approximation for simple and body-centred cubic crystals. In the new formulation of the quasi-chemical method, however, it will be shown that a Legendre transformation helps much in avoiding the mathematical difficulties. (It might be noticed that a similar Legendre transformation is used to essentially the same effect in Fowler's formulation of general statistical mechanics. Cf. Fowler, Statistical Mechanics, second edition, p.188.) The free energy

---
[1] Fowler and Guggenheim, Proc. Roy. Soc. A174, 189(1940).
[2] Bragg and Williams, Proc. Roy. Soc. A145, 699(1934); 151, 540(1935); 152, 231(1935).
[3] Bethe, Proc. Roy. Soc. A150, 552(1935).
[4] Kirkwood, J. Chem. Phys. 6, 70(1938).
[5] Chang, Proc. Camb. Phil. Soc. 35, 265(1939); Kirkwood, J. Chem. Phys. 8, 623(1940); Wang, "Free Energy in the Statistical Theory of Order-Disorder Transformation", Science Report of National Tsing Hua University, Series A, 30-th Anniversary Memorial Number (1941), printed but failed to appear.

is obtained directly as a closed algebraic expression. Its values are given for Bethe's first and second (modified) approximations and for the face-centred alloy Cu_3Au in sections 7 and 8.

To make sure that the quasi-chemical method may actually be used to obtain a series of successively better approximations, we must investigate the free energy in high order approximations and compare it with the partition function of the crystal. This is done in section 5 together with a comparison of the quasi-chemical and Bethe's methods.

Except in the last section we are only concerned with binary alloys with atomic ratio 1:1 forming a (quadratic), simple cubic or body-centred cubic lattice. The generalization of the method to the investigation of alloys with other atomic ratios and forming other types of lattices is easy. In fact, the superior power of the quasi-chemical method appears to be even more fully revealed when a face-centred cubic lattice is treated. This problem is taken at the end of the paper where an approximate free energy expression for Cu_3Au is obtained and its critical phenomena discussed.

2. REFORMULATION OF THE QUASI-CHEMICAL METHOD

Consider a crystal AB. Let $2N$ be the total number of atoms and z the number of nearest neighbours of each. At low temperatures we can distinguish between the so-called α-sites and β-sites for A and B atoms respectively. Denote by Nr the number of A atoms on α-sites. Let

$$1-r = w, \quad r-w = s.$$

The partition function of the crystal is

$$\sum_r p(r,T), \qquad (1)$$

where $p(r,T)$ is equal to $\sum \exp(-E/kT)$ over all possible configurations of the crystal with the given value of r. The average energy over all these configurations is

$$\overline{E}(r,T) = kT^2 \frac{\partial}{\partial T} \log p(r,T) \qquad (2)$$

But evidently,
$$p(r,\infty) = g(r) = \left[\frac{N!}{(Nr)!(Nw)!}\right]^2 \qquad (3)$$

Hence
$$\log p(r,T) = \log g(r) + \int_0^T \frac{1}{kT^2} \overline{E}(r,T) dT. \qquad (4)$$

The problem therefore reduces to one of finding $\overline{E}(r,T)$. Since a direct solution is very difficult we shall try to find an approximate solution by the quasi-chemical method, which is presented below in a form easily generalizable.

There are in the crystal $3N$ nearest pairs of sites α-β. Among these let there be $[q_\alpha, q_\beta]$ with their α-sites occupied by $q_\alpha(=0,1)$ wrong (B) atoms and their β-sites by $q_\beta(=0,1)$ wrong (A) atoms. For given r the following relations hold:

$$[0,1] + [0,0] + [1,1] + [1,0] = 3N,$$
$$[1,1] + [1,0] = 3Nw, \qquad (5)$$

and
$$[0,1] \qquad + [1,1] \qquad = 3Nw.$$

Upon the approximation of neighbour interaction the energy of the crystal may be written as

$$E(r,T) = [0,1]V_{AA} + [0,0]V_{AB} + [1,1]V_{AB} + [1,0]V_{BB}, \qquad (6)$$

where the V's are the interaction energies between a pair of nearest neighbours.

We may give (5) and (6) a different interpretation by imagining $[0,1]$, $[0,0]$, $[1,1]$, $[1,0]$ and V_{AA}, V_{AB}, V_{AB}, V_{BB} to be respectively the numbers and the molecular internal energies of the four different kinds of molecules X_2, X, XYZ, XY of an gaseous assembly. The interpretation of (6) is that the assembly has the same internal (non-kinetic) energy as the crystal at the given value of r. (5) would mean that there are altogether $3N$ X atoms, $3Nw$ Y atoms and $3Nw$ Z atoms in the assembly.

The quasi-chemical method consists in taking the averages $\overline{[0,1]}$,

$\overline{[0,0]}$, $\overline{[1,1]}$, $\overline{[1,0]}$ of the assembly at any temperature as approximately representing the corresponding averages of the crystal at the same temperature. Whether this approximation is good can only be judged for the present from the results it leads to.

A detailed treatment of the problem of a gaseous assembly has been given by Fowler[6]. We are only interested in our assembly of four different kinds of molecules, for which the results may be summarized as:

$$\overline{[0,1]} = \xi \nu e^{-V_{AA}/kT}, \qquad \overline{[0,0]} = \xi e^{-V_{AB}/kT}, \qquad (7)$$
$$\overline{[1,1]} = \xi \mu \nu e^{-V_{AB}/kT}, \qquad \overline{[1,0]} = \xi \mu e^{-V_{BB}/kT},$$

where ξ, μ and ν are to be determined from (6). From (7) we get

$$\overline{[0,0]}\,\overline{[1,1]} / \overline{[0,1]}\,\overline{[1,0]} = X^{-2}, \qquad (8)$$

where
$$X = \exp[-\tfrac{1}{2}(V_{AA} + V_{BB} - 2V_{AB})/kT]. \qquad (9)$$

(8) and (6) together form the starting point of Fowler and Guggenheim's work[1]. The subsequent calculations of $\overline{E}(r,T)$, $p(r,T)$ and the free energy of the crystal are straightforward and will not be repeated here. We shall see later how the free energy can be written down directly without actual integration.

3. GENERALIZATION TO GROUPS OF FOUR SITES

So far we have fixed our attention on the pairs of nearest neighbours in the crystal and have used the quasi-chemical method to obtain the average numbers of the four different kinds of pairs. Now we shall generalize the whole procedure: we shall study all the groups of sites in the crystal of an arbitrarily chosen form in, and classifying these groups according to the way they are occupied by atoms we shall obtain the average number of groups in each class by chemical analogy.

To make this clear let us consider in detail groups of four sites forming squares (as shown) in a quadratic lattice. We classify these groups into $2^4 = 16$ classes

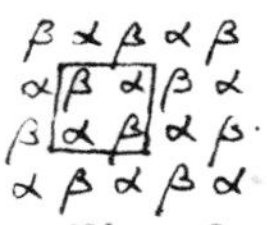

Fig. 1

---
[6] Fowler, Statistical Mechanics, second edition, pp.162-163.

-5-

denoted by $(0,0,0,0)$, $(0,0,0,1)$, ... $(1,1,1,1)$ respectively, so that all groups in the class (q_1,q_2,q_3,q_4) have q_1 wrong atoms in their upper α-sites, q_2 wrong atoms in their lower α-sites, q_3 wrong atoms in their upper β-sites and q_4 wrong atoms in their lower β-sites. The total number of these groups is N. Hence

$$\sum_{q=0}^{1}[q_1,q_2,q_3,q_4]=N, \qquad (10)$$

where $[q_1,q_2,q_3,q_4]$ is an abbreviation for "the number of groups in the class (q_1,q_2,q_3,q_4)". Now the number of all those groups in the crystal with a wrong (B) atom on the upper α-site is just the number of B atoms on the α-sites. Hence

$$\sum_{q} q_i [q_1,q_2,q_3,q_4]=Nw, \qquad i=1,2,3,4. \qquad (11)$$

Let $\chi(q_1,q_2,q_3,q_4)$ be the energy of each group in the class (q_1,q_2,q_3,q_4). It is easy to show that the total energy of the crystal is

$$E(r,T)=\sum_{q}\overline{[q_1,q_2,q_3,q_4]}\chi(q_1,q_2,q_3,q_4). \qquad (12)$$

We may give (10), (11) and (12) an interpretation similar to the one given in section 2 for equations (5) and (6). The same quasi-chemical method used there to obtain (7) leads now to the following averages (approximate) at a given value of r:

$$\overline{[q_1,q_2,q_3,q_4]}=\xi\mu_1^{q_1}\mu_2^{q_2}\mu_3^{q_3}\mu_4^{q_4}e^{-\chi(q_1,q_2,q_3,q_4)/kT} \qquad (13)$$

In this expression the parameters ξ, μ_1, μ_2, μ_3 and μ_4 are to be determined from (10) and (11), which may be written in the form

$$\xi\frac{\partial\varphi}{\partial\xi}=N, \qquad \mu_i\frac{\partial\varphi}{\partial\mu_i}=Nw \quad (i=1,2,3,4), \qquad (14)$$

if we put

$$\varphi(\xi,\mu_1,\mu_2,\mu_3,\mu_4)=\sum_{q}\xi\mu_1^{q_1}\mu_2^{q_2}\mu_3^{q_3}\mu_4^{q_4}e^{-\chi(q_1,q_2,q_3,q_4)/kT} \qquad (15)$$

Or again, in the form

$$\frac{\partial\psi}{\partial\log\xi}=\frac{\partial\psi}{\partial\log\mu_1}=\frac{\partial\psi}{\partial\log\mu_2}=\frac{\partial\psi}{\partial\log\mu_3}=\frac{\partial\psi}{\partial\log\mu_4}=0, \qquad (16)$$

if we put

$$\psi=-N\log\xi-\sum_{i}Nw_i\log\mu_i+\varphi. \qquad (17)$$

It can be shown[7] that ξ and μ_i are uniquely determined by (16) at

---
[7] The proof follows easily (if we put e^{ψ} to be the function Φ) from Lemma 2.42 of Fowler's *Statistical Mechanics*, second edition.

given r and T. Their values at $T=\infty$ are

$$(\xi)_{T=\infty} = Nr^4, \quad (\mu_i)_{T=\infty} = \frac{w}{r}, \quad i=1,2,3,4, \quad (18)$$

as can be verified by substitution into (14).

To calculate the free energy it is necessary first to evaluate the integral in (4). We shall show that this can be done without first solving (14) for ξ and μ_i. For, by (12) and (13) the integrand may be written

$$\frac{1}{kT^2}\overline{E}(r,T) = \frac{1}{kT^2}\sum_{q}\overline{[q_1,q_2,q_3,q_4]}\chi(q_1,q_2,q_3,q_4) = \frac{\partial \varphi}{\partial T}. \quad (19)$$

In the partial differentiation in $\frac{\partial \varphi}{\partial T}$, ξ and μ_i are treated as independent variables. If, however, we regard them as functions (defined by (14)) of r and T, (16) and (17) lead to the following result:

$$\frac{1}{kT^2}\overline{E}(r,T) = \frac{\partial \varphi}{\partial T} = \frac{\partial \psi(r,T)}{\partial T}. \quad (20)$$

Mathematically the change of the independent variables from T, ξ and μ_i to T and r is equivalent to the Legendre transformation

$$\xi, \mu_1, \mu_2, \mu_3, \mu_4 \longrightarrow N, Nw, Nw, Nw, Nw$$

defined by (14). Substituting (20) into (4) we get

$$\log p(r,T) = \log g(r) + \psi(r,T) - \psi(r,\infty), \quad (21)$$

so that the free energy may be written down:

$$F(r,T) = -kT \log p(r,T) = [\log g(r) + \psi(r,T) - \psi(r,\infty)](-kT). \quad (22)$$

The equilibrium value $\bar{r}$ of r is obtained by minimizing F:

$$0 = \frac{\partial F(\bar{r},T)}{\partial \bar{r}} = -kT\left[\frac{d\log g(\bar{r})}{d\bar{r}} + \frac{\partial \psi(\bar{r},T)}{\partial \bar{r}} - \frac{\partial \psi(\bar{r},\infty)}{\partial \bar{r}}\right]. \quad (23)$$

But by (16) and (17)

$$\frac{\partial \psi(r,T)}{\partial r} = \sum_i N \log \mu_i, \quad (24)$$

and by (3)

$$\frac{d\log g(r)}{dr} = 2N \log \frac{w}{r},$$

so that by (18)

$$\sum_i \log \mu_i = -2\log \frac{\bar{w}}{\bar{r}} + \left[\sum_i \log \mu_i\right]_{T=\infty} = 2\log\left(\frac{\bar{w}}{\bar{r}}\right),$$

i.e.

$$\prod_i \mu_i = \left(\frac{\bar{w}}{\bar{r}}\right)^2. \quad (25)$$

It will be shown in the next section that we may put $V_{AA}=V_{BB}$, $V_{AB}=0$, without altering the specific heat of the crystal if $V=\frac{1}{2}(V_{AA}+V_{BB})-V_{AB}$ is will left unchanged. When this is done, ϕ will be symmetrical with respect to μ_1, μ_2, μ_3 and μ_4, and we conclude that all the μ's are equal from the facts that (i) equation (14) has only one set of solution[7], and (ii) if the conclusion is true (14) becomes, with all μ_i put equal to μ,

$$\xi \frac{\partial \phi}{\partial \xi} = N, \qquad \mu \frac{\partial \phi}{\partial \mu} = 4Nw, \qquad (26)$$

which <u>does</u> have[7] a set of solution in ξ and μ. Now ϕ is given by

$$\phi = \xi [1 + 4\mu x^2 + (4\mu^2 x^2 + 2\mu^2 x^4) + 4\mu^3 x^2 + \mu^4], \qquad (27)$$

where x is defined by (9). On eliminating ξ from (26) we obtain

$$(1+S)\mu^4 + (2+4S)x^2\mu^3 + 2Sx^2(x^2+2)\mu^2 + (4S-2)x^2\mu + (S-1) = 0. \qquad (28)$$

The free energy is given by (21) and (18):

$$-\frac{F(r,T)}{2NkT} = r\log r + w\log w - 2w\log \mu + \tfrac{1}{2}\log(1 + 4\mu x^2 + 4\mu^2 x^2 + 2\mu^2 x^4 + 4\mu^3 x^2 + \mu^4) \qquad (29)$$

and the condition of equilibrium by (25):

$$\mu = \sqrt{\frac{1-S}{1+S}}. \qquad (30)$$

To obtain the critical temperature, we expand (28) in powers of s and find after identifying coefficients

$$\log \mu = -\frac{1 + 6x^2 + x^4}{2 + 2x^2} S + KS^3 + \cdots,$$

which is the only real solution for $\log \mu$. Next we expand (30):

$$\log \mu = -S - \tfrac{1}{3}S^3 \cdots.$$

At the critical value x_c of x, these last two equations have a multiple solution at $s=0$. Hence

$$-\frac{1 + 6x_c^2 + x_c^4}{2 + 2x_c^2} = -1,$$

i.e.

$$x_c = (\sqrt{5}-2)^{1/2} = .4858.$$

4. GENERAL FORM OF THE QUASI-CHEMICAL METHOD

Let us now take a group of any size and form. Let it have a α-sites b β-sites and γ pairs of nearest neighbours. The procedures to obtain an approximate expression for the free energy of the crystal follow exactly the same line as in the special case considered in the last section. Equations (13), (14) and (18) are essentially unchanged:

$$\overline{[q_1, q_2, \cdots]} = \zeta \mu_1^{q_1} \mu_2^{q_2} \cdots e^{-\chi/kT}, \tag{31}$$

$$\varphi = \sum_q \zeta \mu_1^{q_1} \mu_2^{q_2} \cdots e^{-\chi/kT}, \quad \zeta \frac{\partial \varphi}{\partial \zeta} = N, \quad \mu_i \frac{\partial \varphi}{\partial \mu_i} = N w, \tag{32}$$

and
$$(\mu_i)_{T=\infty} = \frac{w}{r}. \tag{33}$$

But (12) should be corrected by a factor $\frac{\gamma}{3}$ to account for the duplications in calculating E from the sum of the energies of all the groups in the crystal:

$$E = \frac{3}{\gamma} \sum_q \overline{[q_1, q_2, \cdots]} \chi(q_1, q_2, \cdots) = \frac{3}{\gamma} kT^2 \frac{\partial \varphi}{\partial T}. \tag{34}$$

Hence (22) becomes*

$$F(r,T) = -kT \left[\log g(r) + \frac{3}{\gamma} \varphi(r,T) - \frac{3}{\gamma} \psi(r,\infty) \right], \tag{35}$$

or more explicitly, by (3), (17), and (33):

$$F(r,T) = -\frac{3NkT}{\gamma} \left[\log N + (a+b-\frac{2\gamma}{3})(r\log r + w\log w) - \log \zeta - w \sum_i \log \mu_i \right]. \tag{36}$$

The derivative is
$$\frac{\partial}{\partial r} F(r,T) = -\frac{3NkT}{\gamma} \log\left[(\prod_i \mu_i)(\frac{r}{w})^{a+b-\frac{2\gamma}{3}} \right], \tag{37}$$

so that the condition of equilibrium is

$$\prod_i \mu_i = (\overline{w}/\overline{r})^{a+b-\frac{2\gamma}{3}}. \tag{38}$$

In actual calculations the following points may prove helpful:

(1) <u>The free energy is changed by a constant if V_{AA} and V_{BB} are both replaced by $\frac{1}{2}(V_{AA}+V_{BB}) - V_{AB}$, and V_{AB} by 0.</u> To prove this let j_i be the number of sites in the group neighbouring to the site i. Let χ be changed into χ' by the replacement. It is evident that

$$\chi' - \chi = -\gamma V_{AB} + \frac{V_{AA}-V_{BB}}{2}(\text{no. of B-B pairs} - \text{no. of A-A pairs}),$$

and that
$$\sum_{\alpha\text{-sites}} q_i j_i - \sum_{\mu\text{-sites}} q_i j_i = \text{no. of B-B pairs} - \text{no. of A-A pairs}.$$

Hence
$$\overline{[q_1, q_2, \cdots]} = \zeta \mu_1^{q_1} \mu_2^{q_2} \cdots e^{-\chi/kT} = \zeta' {\mu_1'}^{q_1} {\mu_2'}^{q_2} \cdots e^{-\chi'/kT},$$

if we put
$$\zeta' = \zeta e^{-\gamma V_{AB}/kT}, \quad \mu_i' = \mu_i e^{\pm j_i (V_{AA}-V_{BB})/2kT}$$

where the + sign or the – sign is to be taken according as the site i is an α or a β site. We can now calculate the new free energy and

---

*Care must be taken when the theory is extended to the case when the atomic ratio is not 1:1. The function ψ in (35) must then be replaced by $\frac{1}{2}\psi + \frac{1}{2}\psi'$ where ψ' is the function ψ for the case when the group of interest has the same form as the original one but with α and β sites interchanged.

(-9-)

verify the above statement.

(ii) <u>Sites that are symmetrically situated in the group have equal μ's irrespective of their nature if $V_{AA} = V_{BB}, V_{AB} = 0$</u>. This has already been shown in the last section. Since the most troublesome part of the calculations is the elimination of the parameters, much might be gained by choosing a group with a large number of sites symmetrically situated.

(iii) <u>The free energy is a function of s^2, so that (38) is always satisfied at $\bar{N} = \bar{F} = \tfrac{1}{2}$</u> (i.e. long distance order = 0). The proof is simple when we have already made $V_{AA} = V_{BB}, V_{AB} = 0$, so that an interchange of A and B atoms does not alter the energy. Thus

$$\gamma(q_1, q_2, \ldots) = \gamma(1-q_1, 1-q_2, \ldots).$$

Putting
$$\xi' = \xi \mu_1 \mu_2 \cdots$$

and
$$\mu_i' = 1/\mu_i, \qquad (39)$$

we get
$$\xi \mu_1^{q_1} \mu_2^{q_2} \cdots e^{-\gamma/kT} = \xi' \mu_1'^{1-q_1} \mu_2'^{1-q_2} \cdots e^{-\gamma/kT}.$$

Thus if (32) is satisfied
$$\sum_{q}(1-q_i)\,\xi'\mu_1'^{1-q_1}\mu_2'^{1-q_2}\cdots e^{-\gamma/kT} = \sum_{q}\xi\mu_1^{q_1}\mu_2^{q_2}\cdots e^{-\gamma/kT} - \sum q_i \xi\mu_1^{q_1}\mu_2^{q_2}\cdots e^{-\gamma/kT} = Nr\,;$$

i.e. $\xi', \mu_1', \mu_2', \ldots$ would be the solution of (32) with r substituted for w. Hence by (32) and (17)

$$\Psi(1-r, T) = N - N\log \xi' - \sum_i Nr \log \mu_i' = \Psi(r, T)$$

showing that
$$F(1-r, T) = F(r, T). \qquad (40)$$

(iv) <u>The parameter for a corner site is always given by</u>
$$\epsilon = \frac{1}{1+s}\left(\sqrt{x^2 s^2 + (1-s^2)} - s x\right) \qquad (41)$$

<u>irrespective of the size of the group, if $V_{AA} = V_{BB}, V_{AB} = 0$</u>. By a corner site we mean a site that has only one nearest neighbour in the group. Let ϵ be the selective variable (parameter) of a corner site, and μ_1 that of its only neighbour in the group. If the corner site is dropped, a new group is obtained. We distinguish all quantities referring to this new group by a prime, and obtain at once

-10-

$$\xi' \frac{\partial \phi}{\partial \xi'} = N, \quad \mu_i' \frac{\partial \phi}{\partial \mu_i'} = Nw, \quad i=1,2,\cdots. \quad (42)$$

The sites of the primed group are numbered in the same way as in the unprimed group. Introducing the variable x defined in (9) we may write

$$\phi = \sum_{p,q} \xi \epsilon \mu_1^{q_1} \mu_2^{q_2} \cdots e^{-V/kT}$$

$$= \sum_{p,q} \xi (1+\epsilon x) \mu_2^{q_2} \mu_3^{q_3} \cdots e^{x/kT} + \mu_1 \sum \xi(\epsilon+x) \mu_2^{q_2} \cdots e^{x/kT} \quad (43)$$

Let these two terms be denoted by ϕ_0 and ϕ_1 respectively. Since

$$\phi = N, \quad \mu_1 \frac{\partial \phi}{\partial \mu_1} = Nw,$$

we have
$$\phi_0 = Nr, \quad \phi_1 = Nw. \quad (44)$$

Now $\epsilon \frac{\partial \phi_0}{\partial \epsilon} = \frac{\epsilon x}{1+\epsilon x} \phi_0, \quad \epsilon \frac{\partial \phi_1}{\partial \epsilon} = \frac{\epsilon}{\epsilon + x} \phi_1$.

Hence $\epsilon \frac{\partial \phi}{\partial \epsilon} = Nw$ leads to

$$\frac{\epsilon x}{1+\epsilon x} Nr + \frac{\epsilon}{\epsilon + x} Nw = Nw, \quad (45)$$

or
$$\frac{w}{r} = \frac{\epsilon(\epsilon + x)}{1+\epsilon x}, \quad (46)$$

the solution of which is (41). Thus the two parameters μ and ν in the approximation dicussed in section 2 are equal to ϵ.

(v) <u>The "contribution" to the free energy $\not{p}\not{f}$ from a corner atom is such that, in the notations of (iv),</u>

$$F(r,T) = \frac{\gamma - 1}{\gamma} F'(r,T) + \frac{1}{\gamma} F_0(r,T), \quad (47)$$

where $F_0(r,T)$ is the free energy when $\gamma = 1$, i.e. the free energy in the approximation discussed in section 2. This is proved as follows.

If we put
$$\xi = \xi'' \frac{1}{1+\epsilon x}, \quad \mu_1 = \mu_1'' \frac{1+\epsilon x}{\epsilon + x}, \quad \mu_i = \mu_i'', \quad i \geq 2, \quad (48)$$

it is evident from (43) that ϕ would become a function of $\xi'', \mu_1'', \mu_2'', \cdots$ satisfying the relations

$$\xi'' \frac{\partial \phi}{\partial \xi''} (= \xi \frac{\partial \phi}{\partial \xi}) = N, \quad \mu_i'' \frac{\partial \phi}{\partial \mu_i''} (= \mu_i \frac{\partial \phi}{\partial \mu_i}) = Nw, \quad i=1,2,\cdots. \quad (49)$$

It is also evident that ϕ is the same function of $\xi'', \mu_1'', \mu_2'', \cdots$ as ϕ' is of $\xi', \mu_1', \mu_2', \cdots$. Now (42) has only one[7] set of solution in ξ' and μ_i'. Hence from (49) we infer that $\xi' = \xi''$, $\mu_i' = \mu_i''$. Thus

$$\xi = \xi' \frac{1}{1+\epsilon x}, \quad \mu_1 = \mu_1' \frac{1+\epsilon x}{\epsilon + x}, \quad \mu_i = \mu_i', \quad i \geq 2. \quad (50)$$

(41) and (50) give the parameters μ_i in terms of μ_i'. Inserting them into

(36) we obtain

$$F(r,T) = -NkT\frac{z}{\gamma}\Big[\log N + (a+b-\frac{2\gamma}{z})(r\log r + w\log w) - \log z - w\sum \log \mu' + r\log(1+\epsilon x) + w\log(\epsilon+x) - w\log \epsilon\Big]$$

$$= \frac{\gamma-1}{\gamma}F'(r,T) - NkT\frac{z}{\gamma}\Big[(1-\frac{2}{z})(r\log r + w\log w) + r\log(1+\epsilon x) + w\log\frac{\epsilon+x}{\epsilon}\Big]. \quad (51)$$

If the original (unprimed) group is a pair of nearest neighbours, we have $\gamma=1$, and (51) reduces to the expression for the free energy in the approximation discussed in section 2:

$$F_0(r,T) = -\tfrac{z}{2}NkT\Big[(1-\tfrac{2}{z})(r\log r + w\log w) + r\log(1+\epsilon x) + w\log\frac{\epsilon+x}{\epsilon}\Big]. \quad (52)$$

Inserting this back into (51) we get (47).

5. COMPARISON WITH BETHE'S METHOD

The so-called local grand partition function[1,3,8] in Bethe's approximations with long-distance order is identical in form with our function ϕ when all the "interior sites" in the group have the same parameter μ. For the case of equal concentrations for the two kinds of atoms, which is the case so far considered, this parameter has been put equal to unity by Bethe. Since the different terms of the local grand partition function stand for the probabilities of occurance of the corresponding local groups in the crystal, it is clear that Bethe's method with long-distance order is essentially equivalent to our method plus the assumption that the free energy (35) has a minimum when

$$(\mu)_{\text{interior sites}} = 1. \quad (53)$$

But as we have shown that (38) gives the condition of a minimum of the free energy, the complete* identification of Bethe's and the quasi-chemical methods in any approximation reduces to the mathematical proof of the equivalence of (38) and (53). While this presents no difficulty at all for Bethe's first approximation (section 7), a general proof is by no means easy. We can only satisfy ourselves with the assertion that

---

*"complete" as far as the probabilities of occurance of the local configurations are concerned. The energy calculations are different in the two methods.

[8] Easthope, Proc. Camb. Phil. Soc. 33, 502(1937).

-12-

the two methods are equivalent for large groups, i.e. groups for which
$$a+b-\tfrac{2\gamma}{3} \ll \gamma .$$
This follows from the fact that if (53) is true
$$[(\pi \mu_i)(\tfrac{r}{w})^{a+b-\tfrac{2\gamma}{3}}]^{\tfrac{1}{\gamma}} \cong [\pi(\mu)_{\text{interior sites}}]^{\tfrac{1}{\gamma}} \cong 1 ,$$
so that by (57) $\quad \tfrac{\partial}{\partial r}F(r,T)=0 .$

To see how the equilibrium free energy $F(r,T)$ varies with T in high order approximations, we substitute (38) into (36) and make use of (32):
$$-\tfrac{F}{3NkT}=\tfrac{1}{\gamma}\log(\sum_q \mu_1^{q_1}\mu_2^{q_2}\cdots e^{-\chi/kT})+\tfrac{1}{\gamma}(a+b-\tfrac{2\gamma}{3})\log \overline{r} .$$
The last term is very small for large groups, so that by (53)
$$-\tfrac{F}{3NkT}=\tfrac{1}{\gamma}\log(\sum e^{-\chi/kT}) .$$

6. THE NON-INTERFERENCE OF LOCAL CONFIGURATIONS

Let us return to the fundamental assumption of the quasi-chemical method, i.e. to (31) which gives the average numbers of the different local configurations (so far called groups) in the crystal. ~~Distinguishing all quantities~~ This equation expresses the exact distribution law of an assembly of molecules (cf. the example in section 2) which has an energy $\tfrac{\gamma}{3}$ times as large as the crystal. Distinguishing all quantities referring to the assembly of molecules by a subscript m, we get
$$F(r,T)+kT\log g(r)=\tfrac{3}{\gamma}[F_m(r,T)+kT\log g_m(r)],$$
which is obtained from (4). But if H is the number of arrangements in the crystal lattice having the given values of $[q_1, q_2, \cdots]$,
$$F(r,T)=-kT\log \overline{H}+\overline{E} . \tag{54}$$

Thus $\quad \log \tfrac{\overline{H}}{g(r)}=\tfrac{3}{\gamma}\log \tfrac{\overline{H}_m}{g_m(r)} .$

But* $\quad H_m = \tfrac{N!}{\prod_q [q_1,q_2,\cdots]!} ,\tag{55}$

hence dropping the bar we get
$$H =h(r)\{\tfrac{N!}{\prod_q [q_1,q_2,\cdots]!}\}^{3/\gamma}, \tag{56}$$
where $\quad h(r)=g(r)/\{g_m(r)\}^{3/\gamma} . \tag{57}$

-----
*Fowler, Statistical Mechanics, second edition, sections 2.6 and 5.11.

-13-

Equation (56) has been referred to in Fowler and Guggenheim's paper[1] as the mathematical expression of the "hypothesis of the non-interference of local configurations", because when $\frac{\gamma}{3}=1$, the number of arrangements in the crystal consistent with the distribution law $[q_1, q_2, \cdots]$ for the groups of sites is, except for the factor h(r), equal to

$$H_m = \frac{N!}{\prod [q_1, q_2, \cdots]!},$$

which is the number of arrangements in the crystal for the given values of $[q_1, q_2, \cdots]$ if the N groups in the crystal are <u>imagined</u> to be <u>separated</u> and are filled <u>independently</u> with atoms. The term "non-interference" comes from the fact that actually the N groups are <u>not separated</u> but are <u>interlocked</u> and <u>cannot be filled independently</u> with atoms, i.e. they "interfere" with each other.

To find the value of $g_m(r)$ we notice that by definition $g_m = \Sigma H_m$. But ΣH_m is the number of arrangements in the N separated groups considered above if they are to be so filled with atoms that Nw of them have wrong atoms on the sites i, $i=1, 2, \cdots$. Among the N sites i of the N groups $\frac{N!}{(Nr)!(Nw)!}$ different arrangements are possible. Hence*

$$g_m = \Sigma H_m = \Sigma \frac{N!}{\prod[q_1, q_2, \cdots]!} = \left[\frac{N!}{(Nr)!(Nw)!}\right]^{a+b} \quad (58)$$

Thus
$$h(r) = \left[\frac{N!}{(Nr)!(Nw)!}\right]^{2(a+b)\frac{3}{\gamma}} \quad (59)$$

The free energy of the crystal may be obtained from (54), (56) and (59):

$$F(r,T) = \overline{E} - \frac{3NkT}{\gamma}\left\{(a+b-\frac{2\gamma}{3})(r\log r + w\log w) + \log N - \frac{1}{N}\Sigma[q_1, q_2, \cdots]\log[q_1, q_2, \cdots]\right\}$$

which has been obtained above by integration.

------

*It might be mentioned in passing that for the special case considered in section 2, (58) gives directly the value of the sum Σ_x in equation (8.5) of Fowler and Guggenheim's paper if their r and q are equal. The generalization to the case r≠q is however easy. The result is

$$\Sigma_x \frac{[3N]!}{[3N(r-x)]![3Nx]![3N(1-r-q+x)]![3N(q-x)]!} = \frac{[3N]!}{[3Nr]![3N(1-r)]!} \cdot \frac{[3N]!}{[3Nq]![3N(1-q)]!} \quad (60)$$

which is exact. The value of $\log \Sigma_x$ given by (60) reduces to the approximate expression that Fowler and Guggenheim obtained by identifying Σ_x with its maximum term when N is large.

-14-

7. SPECIAL CONSIDERATIONS CONCERNING BETHE'S FIRST AND SECOND APPROXIMATIONS

(i) <u>First Approximation</u>. If an α-site together with its γ nearest neighbours are taken as our group of interest, all the sites except the central one are corner sites. Hence their selective variables are all equal to the value of ϵ given in (41). By successive applications of (47) we see that the free energy is exactly $F_c(r,T)$, a fact which has already been pointed out by Fowler and Guggenheim[1]. The selective variable of the central site is given by successive applications of (50)

$$\lambda = \frac{w}{r}\left(\frac{1+\epsilon x}{\epsilon + x}\right)^{\gamma} \quad . \tag{61}$$

The factor w/r is the selective variable for the central site when it alone forms the group. The equilibrium condition (38) becomes

$$\lambda \epsilon^{\gamma} = (w/r)^{\gamma - 1} \quad .$$

But by (61) and (45), $\qquad \lambda = (w/r)^{1-\gamma} \epsilon^{\gamma} \quad ,$

Hence at equilibrium $\qquad \lambda = 1 \quad . \tag{62}$

Thus the approximation is completely equivalent to Bethe's first approximation, as already mentioned in section 5.

(ii) <u>Second Approximation</u>. Now consider the group of sites occuring in Bethe's second approximation[3]. According to section 4, (iv), the selective variables for the corner sites in the second shell are all equal to ϵ, which is given by (41). But in Bethe's original calculations, the selective variables for the corner sites and the medium sites are made equal, and are found to be different from ϵ. Thus if we use his original method, equation (32) can not be satisfied.(In other words, the probabilities of occurence of wrong atoms in the corner and the medium sites would be unequal.)

For simplicity we shall drop the corner sites and take as our group of interest the central site, the first shell sites and the medium

-15-

sites; with selective variables μ, ν and λ respectively. (The contribution ~~to the free energy by~~ by the corner sites can be included in the free energy by simple addition as shown in section 4 (v).) With the notations n, m and g_{nm} of Bethe[3] we find

$$\Phi = 3 \sum_n (x^n + \mu x^{3-n}) P_n(x,\nu,\lambda), \quad (63)$$

where $P_n(x,\nu,\lambda) = \nu^n \sum_m g_{nm}[(1+\lambda)x]^m (x^2+\lambda)^{(\frac{3}{2}-1)n-\frac{m}{2}} (1+\lambda x^2)^{(\frac{3}{2}-1)(3-n)-\frac{m}{2}}$.

After eliminating ξ and μ, (52) becomes

$$3w = \frac{2 \sum_{l,n} x^{l-n}(r P_n \lambda \frac{\partial}{\partial \lambda} P_l + w P_l \lambda \frac{\partial}{\partial \lambda} P_n)}{(3-2)(\sum_n x^n P_n)(\sum_n x^{-n} P_n)} = \frac{\sum_{l,n} x^{l-n}(r P_n \nu \frac{\partial}{\partial \nu} P_l + w P_l \nu \frac{\partial}{\partial \nu} P_n)}{(\sum_n x^n P_n)(\sum_n x^{-n} P_n)}. \quad (64)$$

The free energy is obtained from (36):

$$F(r,T) = -\frac{NkT}{3-1}\left[\frac{1}{2}(3^2-43+4)(r\log r + w\log w) + r\log\sum_n x^n P_n + w\log\sum_n x^{-n} P_n - 3w\log\nu - w3(\frac{3}{2}-1)\log\lambda\right]. \quad (65)$$

8. APPLICATION TO THE CRYSTAL Cu₃Au

For the face-centred crystal Cu₃Au, we may of course follow Peierls[9] and take as our group a central site together with its twelve first shell neighbours. The free energy expression would then contain seven selective variables*, four of which can be eliminated. The resultant expression is very cumbersome and numerical calculations would be laborious. We therefore make a simpler approximation: the group is taken to be four nearest neighbours forming a tetrahedron. A little geometrical consideration assures us that all such tetrahedrons contain an

Fig.2

α-site (for gold atoms) and three β-sites (for copper atoms), an interesting conclusion showing that the tetrahedron might be regarded as a sort of "molecular" structure in a face-centred lattice with atomic ratio 1:3. Our approximation may thus be reasonably expected to reveal the more important features of the order-disorder transformation in such

---
*For α-centred groups, three, and for β-centred, four, parameters are necessary. Both these two kinds of groups must be considered because otherwise the energy of the crystal cannot be easily obtained from the energy of the groups in the crystal.

[9] Peierls, Proc. Roy. Soc. London A154, 207(1936).

-16-

alloys.

Let μ and ν be the parameters (for wrong atoms) of the β-sites and the α-sites respectively. Let there be altogether $4N$ atoms. It is easy to see that there are $8N$ groups in the crystal. When Nw atoms on the α-sites are wrong, the equations determining the parameters are

$$8N = \phi = \xi[x^3 + 3x^2\mu + 3x^3\mu^2 + x^6\mu^3 + \nu(x^6 + 3x^3\mu + 3x^2\mu^2 + x^3\mu^3)], \tag{66a}$$

$$8Nw = \nu\frac{\partial \phi}{\partial \nu} = \xi\nu(x^6 + 3x^3\mu + 3x^2\mu^2 + x^3\mu^3), \tag{66b}$$

and $\quad 8N(\frac{w}{3}) + 8N(\frac{w}{3}) + 8N(\frac{w}{3}) = \mu\frac{\partial \phi}{\partial \mu} = 3x^2\mu\xi[1 + 2x\mu + x^4\mu^2 + \nu(x + 2\mu + x\mu^2)], \tag{66c}$

where x is defined by (9). The energy of the crystal is (cf. (34)),

$$E = \frac{1}{2}kT^2\frac{\partial \phi}{\partial T} + \text{constant}; \tag{67}$$

so that the free energy becomes (cf. (35))

$$F(w,T) = -kT\left[\log g(w) + \frac{1}{2}(\phi - 8N\log\xi - 8Nw\log\nu - 8Nw\log\mu)\Big|_{T=\infty}^{T}\right].$$

But $\quad \log g(w) = -N\left\{(1-w)\log(1-w) + w\log w + w\log\frac{w}{3} + (3-w)\log(3-w)/3\right\},$

and at $T = \infty \quad \nu = \frac{w}{1-w}, \quad \mu = \frac{w}{3-w}, \quad \xi = 8N(1-w)(1-\frac{w}{3})^3.$

Hence $\quad -\frac{F(w,T)}{NkT} = -9\log 3 + 4\log 8N + 6w\log w + 3(1-w)\log(1-w) + 3(3-w)\log(3-w)$
$\qquad\qquad -4\log\xi - 4w\log\mu - 4w\log\nu. \tag{68}$

Since ξ and ν can be very easily solved from (66), numerical calculations are quite simple. The equilibrium value of w is given by (cf. (37) and (38))

$$0 = 3\log\frac{(1-\bar{w})(3-\bar{w})}{\bar{w}^2} + 4\log\mu\nu. \tag{69}$$

This is always satisfied at $\bar{w} = \frac{3}{4}$.* Actual calculation shows that the absolute minimum of the free energy is or is not at $\bar{w} = \frac{3}{4}$ according as $x \geq .2965$ or $x < .2965$. The values of the free energy is plotted in Fig.3. From the form of the graph it is seen that the crystal has a critical temperature at which the long-distance order and (hence) the energy are discontinuous. The critical temperature T_c and the latent heat Q are found to be

---
*This is not evident from (69) directly. But if we divide the whole crystal into four sublattices which are all simple cubic and introduce a w for each sublattice so that Nw_i is the number of A atoms on the i-th sublattice (i=1,2,3,4), it is obvious that the free energy is symmetrical in the w's. From this we infer that (69) is satisfied at $\bar{w} = \frac{3}{4}$.

-17-

$$T_c = .8228 \frac{1}{k}[\tfrac{1}{2}(V_{AA} + V_{BB}) - V_{AB}] \quad , \quad Q = .6624 N[\tfrac{1}{2}(V_{AA} + V_{BB}) - V_{AB}].$$

In terms of the total energy change from $T=0$ to $T=\infty$:

$$E_c = 3N[\tfrac{1}{2}(V_{AA} + V_{BB}) - V_{AB}]$$

these quantities become

$T_c = 1.097 E_c/R$, ($T_c = 2.19 E_c/R$ in Bragg-Williams' approximation and
$T_c \cong 1.5 E_c/R$ in Peierls' approximation.)

$Q = .2941 E_c$, ($Q = .218 E_c$ in Bragg-Williams' approximation and
$Q \cong .36 E_c$ in Peierls' approximation.)

where R stands for $4Nk$.

It will be noticed that due to the lack of a free energy expression Peierls[9] did not give the exact values of these quantities.

In conclusion, the author wishes to express his thanks to Prof. J. S. Wang for valuable criticism ~~advice~~ and advice.

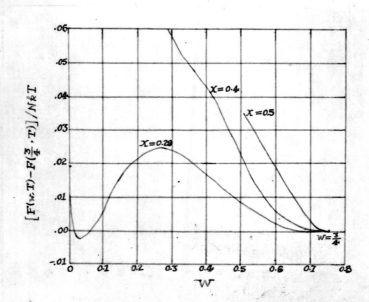

Fig. 3

On the Angular Distribution in Nuclear Reactions and Coincidence Measurements

1948 thesis for PhD degree

THE UNIVERSITY OF CHICAGO

ON THE ANGULAR DISTRIBUTION IN NUCLEAR REACTIONS
AND COINCIDENCE MEASUREMENTS

A DISSERTATION SUBMITTED TO
THE FACULTY OF THE DIVISION OF THE PHYSICAL SCIENCES
IN CANDIDACY FOR THE DEGREE OF
DOCTOR OF PHILOSOPHY

DEPARTMENT OF PHYSICS

BY
CHEN NING YANG

CHICAGO, ILLINOIS
JUNE, 1948

INTRODUCTION

In the calculation of the angular distribution in nuclear reactions and of the angular correlation in processes involving β- and γ-decay it often happens that many terms cancel out at the end of a laborious computation. The consistency of the occurrence of such cancellation leads one to suspect that some general reasons quite independent of the particular form of interaction are at work. In this paper we shall show that this is indeed the case. In fact the general form of the angular distribution in many cases can be obtained directly from the theorems derived in this paper.

For nuclear reactions between spinless particles the existence of a limitation on the complexity of the angular distribution for fixed orbital angular momentum of the incoming particles is well known. That the same result holds with the spin taken into consideration (for unpolarized incoming beam) was first pointed out by Critchfield and Teller.[1] A proof of this statement was recently given by Eisner, Sachs, and Wolfenstein.[2] We shall in this paper formulate a new proof that lends itself easily to generalization to the case in which the particles involved have relativistic velocities.

It will be shown in general that in studying the angular correlation between two particles, as long as one of them has a wave length long compared to the size of the nucleus, the process can be classified into different orders and for a process of given order the general form of the angular correlation is essentially known. In case both of the particles have long wave lengths, particularly simple conclusions may be reached, as in the case of β-neutrino correlation in β-decay.

Experimentally the angular correlations β–neutrino and γ–γ have been studied by many authors. Various calculations of these correlations based on different kinds of interactions have also been made. These will be separately discussed in the different sections.

NUCLEAR REACTION

Consider the following reaction

$$A + P \to B + Q \tag{1}$$

and suppose both the target nucleus A and the bombarding beam of particles P are unpolarized. The complexity of the angular distribution of the outgoing particles is limited by the following theorem: If only incoming waves of orbital angular momentum L contribute appreciably to the reaction, the angular distribution of the outgoing particles

---

[1] C. L. Critchfield and E. Teller, Phys. Rev. 60, 10 (1941).
[2] E. Eisner and R. G. Sachs, Phys. Rev. 72, 680 (1947); L. Wolfenstein and R. G. Sachs, 73, 528 (1948).

in the centre of mass system is an even polynomial of $\cos\theta$ with maximum exponent not higher than $2L$. Here θ is the angle between the incoming and the outgoing particles in the centre-of-mass system of reference.

To prove this let us consider the collision between two particles A and P with definite (a and p) components of spin along the z-axis, and definite total and z-component relative orbital angular momenta L and m. (We use the center of mass system throughout.) The incoming wave function is, at large distances r_{AP} between A and P:

$$\frac{1}{r_{AP}}\sin(k_{AP}r_{AP} - \frac{1}{2}L\pi)\psi_A^a\psi_P^p Y_{Lm}(\theta_P,\phi_P) \tag{2}$$

where ψ_A^a, ψ_P^p are normalized internal wave functions of particles A and P; θ_P, ϕ_P describe the direction of motion of the particle P; and $Y_{Lm}(\theta,\phi)$ is the normalized spherical harmonic of order Lm.

The asymptotic behavior of the wave function at large values of r_{BQ} is of the form

$$\frac{1}{r_{BQ}}\exp(ik_{BQ}r_{BQ})\sum_{bq}\psi_B^b\psi_Q^q f_{bq}^{apm}(\theta_Q,\phi_Q). \tag{3}$$

In reaction (1), if we choose as the z-axis the direction of motion of particle P, it is clear that when the incoming wave is expanded into partial waves with definite total and z-component orbital angular momenta L and m only terms with $m = 0$ occur. Under the assumption stated in the theorem we can neglect all terms except the spherical harmonic Y_{L0}. The differential cross-section of reaction (1) is therefore

$$d\sigma = (\text{constant})\,d\Omega_Q\sum_{bq}\left|f_{bq}^{ap0}(\theta_Q,\phi_Q)\right|^2. \tag{4}$$

For unpolarized incoming particles we get

$$d\sigma = (\text{constant})\,d\Omega_Q\sum_{apbq}\left|f_{bq}^{ap0}(\theta,\phi)\right|^2. \tag{5}$$

The requirement of invariance under rotation will now be introduced. Consider a new coordinate system (primed system) obtained from the old by a rotation of the coordinate axis. Let $(m'|m)^{(L)}$ be the matrix element [3] of the irreducible representation $\mathfrak{D}^L$ of the three dimensional rotation group. We have

$$Y_{Lm'}(\theta',\phi') = \sum_m (m'|m)^{(L)} Y_{Lm}(\theta,\phi),$$
$$\psi_A'^{a'} = \sum_a (a'|a)^{(S_A)} \psi_A^a, \tag{6}$$

---
[3] E. Wigner, Gruppentheorie und ihre Anwendung auf die Quantenmechanik der Atomspektren (1931), 180.

where S_A = spin of particle A, which may be an integer or a half odd integer. By $\psi_A^{\prime a'}$ is meant the function $\psi_A^{a'}$ of the <u>primed</u> (internal) coordinates. The proof of our theorem consists in showing that (i) the superposition principle requires that f be transformed according to $\mathfrak{D}^{S_A} \times \mathfrak{D}^{S_P} \times \mathfrak{D}^L \times \mathfrak{D}^{S_B^*} \times \mathfrak{D}^{S_Q^*}$ and (ii) the expression

$$I^{mm'}(\theta,\phi) = \sum_{apbq} [f_{bq}^{apm}(\theta,\phi)]^* f_{bq}^{apm'}(\theta,\phi) \tag{7}$$

transforms according to $\mathfrak{D}^{L^*} \times \mathfrak{D}^L = \mathfrak{D}^{2L} + \mathfrak{D}^{2L-2} + \cdots$. No non-vanishing linear combinations of (7) transform according to $\mathfrak{D}^{2L-1}$, $\mathfrak{D}^{2L-3}$, $\cdots$, or $\mathfrak{D}^1$.

(i) Consider the following incoming wave

$$\frac{1}{r_{AP}} \sin(k_{AP} r_{AP} - \frac{1}{2} L\pi) \psi_A^{\prime a'} \psi_P^{\prime p'} Y_{Lm'}(\theta_P', \phi_P'). \tag{2'}$$

To an observer in the primed coordinate system this has exactly the same form as (2). Hence the outgoing wave must be

$$\frac{1}{r_{BQ}} \exp(ik_{BQ} r_{BQ}) \sum_{b'q'} \psi_B^{\prime b'} \psi_Q^{\prime q'} f_{b'q'}^{a'p'm'}(\theta_Q', \phi_Q'). \tag{3'}$$

Notice that we use the <u>same</u> f instead of an f', because there is <u>no physically observable distinction</u> between the two coordinate systems. Using (5) one can express (2') as a superposition of waves (2)

$$\sum_{apm} (a'|a)(p'|p)(m'|m) \left[\frac{1}{r_{AP}} \sin(k_{AP} r_{AP} - \frac{1}{2} L\pi) \psi_A^a \psi_P^p Y_{Lm}(\theta_P, \phi_P) \right].$$

Here we have omitted the superscripts S_A, S_P, L from $(a'|a)^{(S_A)}$, $(p'|p)^{(S_P)}$, $(m'|m)^{(L)}$ for simplicity. The outgoing wave must therefore be a corresponding superposition of waves (3) with the same coefficients:

$$\sum_{apm} (a'|a)(p'|p)(m'|m) \left[\frac{1}{r_{BQ}} \exp(ik_{BQ} r_{BQ}) \sum_{b,q} \psi_B^b \psi_Q^q f_{bq}^{apm}(\theta_Q, \phi_Q) \right].$$

Equating this to (3') and using (6) to express $\psi_B^{\prime b'}$, $\psi_Q^{\prime q'}$ in terms of ψ_B^b, ψ_Q^q we get finally by identifying the coefficient of $\psi_B^b \psi_Q^q$:

$$\sum_{apm} (a'|a)(p'|p)(m'|m) f_{bq}^{apm}(\theta,\phi) = \sum_{b'q'} (b'|b)(q'|q) f_{b'q'}^{a'p'm'}(\theta',\phi').$$

This reduces to the following form

$$f_{b'q'}^{a'p'm'}(\theta',\phi') = \sum_{apmbq} (a'|a)(p'|p)(m'|m)(b'|b)^*(q'|q)^* f_{bq}^{apm}(\theta,\phi) \tag{8}$$

through the orthogonality relations

$$\sum_m (m'|m)(m''|m) = \delta_{m'm''}. \tag{9}$$

Equation (8) expresses the transformation property of f.

(ii) To obtain the transformation property of expression (4) we investigate the behavior of expression (7) under rotation. By (8) and (9)

$$I^{m''m'}(\theta', \phi') = \sum_{mm'''}\sum_{abpq}(m''|m''')^*(m'|m)\left[f_{bq}^{apm'''}(\theta,\phi)\right]^* f_{bq}^{apm}(\theta,\phi)$$

$$= \sum_{mm'''}(m''|m''')^*(m'|m) I^{m'''m}(\theta,\phi). \tag{10}$$

Now the differential cross-section is proportional to I^{00}. If we put in this expression $m'' = m' = 0$ and take the rotation from the unprimed to the primed coordinate system to be a rotation around the z-axis by an angle ξ we have $\theta' = \theta$ and $\phi' = \phi + \xi$. Since then $(0|m) = \delta_{m0}$ it is evident from (10) that

$$I^{00}(\theta, \phi+\xi) = I^{00}(\theta,\phi)$$

showing that I^{00} is independent of ϕ. To study its dependence on θ we put in (10) $m' = m'' = 0$. It is well known[4] that if $\theta = \phi = 0$

$$(0|m) = Y_{L,-m}(\theta',\phi').$$

Hence (10) becomes

$$I^{00}(\theta',\phi') = \sum_{mm'''} Y_{L,-m'''}^*(\theta',\phi') Y_{L,-m}(\theta',\phi') I^{m'''m}(0,0).$$

On application of the reduction theorem[4] of products of spherical harmonics this leads directly to our theorem.

If instead of a rotation we had chosen an inversion of the coordinates it is evident that (8) would become

$$f_{bq}^{apm}(\pi - \theta, \pi + \varphi) = \mathscr{P}_A \mathscr{P}_P \mathscr{P}_B \mathscr{P}_Q (-1)^L f_{bq}^{apm}(\theta,\phi) \tag{11}$$

where $\mathscr{P}_A$, $\mathscr{P}_P$ etc. are the intrinsic parities of the nuclei. This shows that

$$|f(\pi-\theta, \pi+\phi)|^2 = |f(\theta,\phi)|^2$$

and it follows that the angular dependence must be an even function of $\cos\theta$, a fact that is already established by (10). Equation (11) further shows that any odd power of

---

[4] H. Bethe, Handbuch der Physik, Band 24/1, Kap. 3, Section 65.

$\cos\theta$ in the angular dependence must come from an interference term between orbital wave functions of opposite parity.

The symmetry requirements of the wave function under interchanges of the nucleons do not[5] in general lead to any new conclusions about the properties of f. However, in the special case in which the two incoming particles or the two outgoing particles are identical more detailed consideration is necessary. An example of such a case is the reaction

$$\text{Li}^7 + \text{H}^1 \to \text{He}^4 + \text{He}^4.$$

Since the outgoing particles are spinless and satisfy Bose-Einstein statistics and since Li^7 has an odd parity, the value of L must be odd in order to have a balance of parity. This means that $f = 0$ unless L is odd. At low energies, therefore, the effective orbital angular momentum of the incoming particles is 1.

Another example is the $\text{D}^2 + \text{D}^2$ reaction:

$$\text{D}^2 + \text{D}^2 \to \text{N}^1 + \text{He}^3,$$
$$\text{D}^2 + \text{D}^2 \to \text{H}^1 + \text{H}^3.$$

This reaction has recently been considered theoretically by Konopinski and Teller.[6] Due to the symmetry nature of the deuterons it is no longer convenient to specify the spin of the two incoming particles separately. Instead we should group the nine possible incoming states into a quintet, a triplet, and a singlet. The space wave functions for the quintet and the singlet states are symmetrical with respect to the exchange of the two deuterons and those for the triplet states are antisymmetrical. Strictly speaking the proof of our theorem does not apply to such a case where the space wave function depends on the orientation of the spins of the particles. But since all the states in the same multiplet have the same a priori probability it is evident that the difference of the space wave function for the different multiplets does not affect the validity of our theorem.

The Coulomb field affects the waves of different orbital angular momenta in such a way as to favor those with higher angular momenta at low energies.[7] This accounts for the reason why at bombarding energies as low as 20 keV. The angular distribution in the D + D reaction is not symmetrical spherically.[6,8] We shall not go into this point in any further details here.

---

[5]To understand this it is best to introduce the idea of channels in the configuration space, which was first discussed by G. Breit, Phys. Rev. 58, 1068 (1940); J. A. Wheeler, Phys. Rev. 52, 1107 (1937). An interchange of the nucleons in general results in an interchange of the channels, except for the case when either A and P or B and Q are identical.

[6]E. J. Konopinski and E. Teller, Phys. Rev. 73, 822 (1948).

[7]H. Bethe and E. J. Konopinski, Phys. Rev. 54, 130 (1938).

[8]E. Bretscher, A. P. French, and F. G. P. Seidl, Phys. Rev. 73, 815 (1948).

We conclude this section by stating a variation of the theorem proved above: When contributions from incoming waves with orbital angular momenta L are neglected, the angular distribution in reaction (1) in the centre of mass system is a polynomial of $\cos\theta$ with maximum exponent not higher than $2L$. This holds even if the contributing compound nuclear states have angular momenta $> L$.

It will be noticed that when both even and old values of the orbital angular momenta in the incoming beam are effective in producing the reaction the angular distribution contains odd powers of $\cos\theta$. This, however, will not happen when either (i) the reaction goes through a single compound nuclear state (e.g. near a strong resonance level); or (ii) symmetry requirements exclude even (or odd) L values as in the $\text{Li}^7 + \text{H}^1 \to \text{He}^4 + \text{He}^4$ reaction discussed above.

RELATIVISTIC CASE

We shall in this section generalize the result of the last section to the case when the particle P is an electron and has relativistic velocities. (The nuclei A, B, and Q are still supposed to be nonrelativistic.) No such process has been experimentally realized. We shall, however, discuss it to illustrate our method. It will be proved that if only partial waves of orbital angular momentum L in the electron wave function contribute to the reaction the angular correlation is a polynomial of $\cos\theta$ with maximum exponent not higher than $2L+1$.

Instead of the stationary picture used in the last section we shall here use a non-stationary description of the process. The electron wave function at time $t=0$ is a product of a spin wave function with four components and a space wave function e^{ikz}. The spin of the electron along the z-axis is a constant of motion and is denoted by $p\,(=\pm\frac{1}{2})$. If we expand the space wave function into partial waves of definite orbital angular momenta L, the first term ($L=0$) would give rise to allowed transitions, the second term ($L=1$) first forbidden transitions etc. To study the angular correlation arising from the contribution of the partial wave of orbital angular momentum L we need to decompose it again into normalized wave $\psi_{LJ\mathscr{P}m}$ of definite L, J (total angular momentum of the electron), $\mathscr{P}$ (parity), and m (z-component of $\vec{J}$)*. The advantage of using these ψ's is that they have simple transformation properties under rotation. The possible values of J are $L\pm\frac{1}{2}$. Under the assumption that we are only considering the contribution from a definite L value the wave function at $t=0$ can be replaced by

$$\sum_{\mathscr{P}=\pm 1}\sum_{J=L\pm 1/2} \alpha_{LJ\mathscr{P}p}\,\psi_{LJ\mathscr{P}p}. \tag{12}$$

We have put $m=p$ because the z-component of the orbital angular momentum is zero.

$\mathscr{P}$ ranges over 1 and -1.*

Let us now first study the reaction arising from the electron wave $\psi_{LJ\mathscr{P}m}$. Starting at $t = 0$ with $\psi_{LJ\mathscr{P}m}$ and nucleus A with a definite value a for the z-component of spin, we shall denote by $f_{bq}^{LJ\mathscr{P}am}(\theta_Q, \phi_Q)$ the probability amplitude at any later time $t > 0$ for that outgoing state in which the z-component of spin of the particles B and Q are b and q, and in which the momentum of Q is in the direction θ_Q, ϕ_Q. The absolute value of the outgoing momentum (which is not fixed because the energy is not necessarily conserved when t is small) should also enter the function f as an independent variable, but has been omitted for simplicity of writing.

Now the probability amplitudes are additive when we superpose states. Since under a rotation the different waves $\psi_{LJ\mathscr{P}m}$ with the same $L, J, \mathscr{P}$ values combine linearly, the argument which led to (8) in the last section would now lead to

$$f_{b'q'}^{LJ\mathscr{P}a'm'}(\theta', \phi') = \sum_{ambq} (a'|a)(b'|b)^*(m'|m)(q'|q)^* f_{bq}^{LJ\mathscr{P}am}(\theta, \phi). \tag{13}$$

Returning now to the wave (12) at $t = 0$ we see that the differential cross-section is proportional to

$$d\Omega_Q \left| \sum_{\mathscr{P}J} \alpha_{LJ\mathscr{P}p} f_{bq}^{LJ\mathscr{P}ap}(\theta, \phi) \right|^2.$$

This will have to be summed over a, b, p, and q. Since the coefficients in (12′) are independent of a, b, and q the final expression is

$$d\Omega_Q \sum_{J\mathscr{P}J'\mathscr{P}'p} \alpha_{LJ\mathscr{P}p} \alpha_{LJ'\mathscr{P}'p}^* \left[\sum_{abq} \left\{ f_{bq}^{LJ'\mathscr{P}'ap}(\theta, \phi) \right\}^* f_{bq}^{LJ\mathscr{P}ap}(\theta, \phi) \right]. \tag{14}$$

By (13) the individual terms under the summation sign $\sum_{abq}$ transform under a rotation according to $\mathfrak{D}^{S_A} \times \mathfrak{D}^{S_A^*} \times \mathfrak{D}^{S_B} \times \mathfrak{D}^{S_B^*} \times \mathfrak{D}^{S_Q} \times \mathfrak{D}^{S_Q^*} \times \mathfrak{D}^{J} \times \mathfrak{D}^{J'^*}$. But after the summation over a, b, and q is carried out, the sum transforms more simply according to $\mathfrak{D}^{J} \times \mathfrak{D}^{J'^*} = \mathfrak{D}^{J+J'} + \mathfrak{D}^{J+J'-1} + \cdots$. This means that the expression in the square bracket in (14) is a sum of spherical harmonics of order $\mathscr{L}\mathfrak{M}$ with $\mathscr{L} \leq J + J'$. But both J and J' are $\leq L + \frac{1}{2}$. The theorem stated at the beginning of this section follows immediately.

If we introduce the requirement of invariance under inversion (14) shows that those terms with $\mathscr{P}'\mathscr{P} =$ even give rise to angular correlation functions that are even under the transformation $\theta \to \pi - \theta$; and those with $\mathscr{P}'\mathscr{P} =$ odd give rise to odd angular correlation functions. A consequence of this is the following. If the velocity v of the

---

*The parity can be either 1 or -1 for any given L, J, and m. However, for slow electrons the amplitude of waves with $\mathscr{P} = -(-1)^L$ is very small. Cf. the end of this section.

electron is small compared to the velocity of light c, and if the spin wave function of the electron is expanded in powers of v/c, the first term, i.e. the term that does not vanish as $v \to 0$, is invariant under an inversion. This term would therefore give rise to terms with $\mathscr{P} = (-1)^L$. The opposite parity first appears in the next term of the expansion and is proportional to v/c. Hence those terms in (14) with $\mathscr{P}'\mathscr{P} =$ odd contain a factor v/c. Thus the odd powers of $\cos\theta$ in the angular correlation have coefficients smaller than the even powers by a factor of v/c.

β–NEUTRINO CORRELATION

In β-decay we have the particularly simple situation in which both the electron and the neutrino have wave lengths long compared to the dimension of the nucleus. The argument of the last section can now be applied to both these particles and we can prove that the angular correlation between the electron and the neutrino emitted in a β-decay is a polynomial of $\cos\theta$ up to a maximum exponent $K + 1$, where[9] $K = 0$ for allowed transitions, $K = 1$ for first forbidden transitions etc.[9]

The idea of the proof is that for first forbidden transitions one has either $L = 1$ for the electron and $L_1 = 0$ for the neutrino or $L = 0$ for the electron and $L_1 = 1$ for the neutrino. The waves $L = 1$ and $L_1 = 1$ occur together only in second forbidden processes. Now the intensity produced by the $L = 1$, $L_1 = 0$. Waves has an angular correlation function that goes up to cos to the first power, according to the theorem of the last section. Similarly fixing our attention on the neutrino wave function we can draw the same conclusion about the $L = 0$, $L_1 = 1$ waves. The interference term of the $L = 1$, $L_1 = 0$ waves with the $L = 0$, $L_1 = 1$ waves, however, gives an angular distribution that contains $\cos^2\theta$, which is the highest power of $\cos\theta$ possible for this case.

The proof is as follows. Consider the β-decay

$$A \to B + e + \nu.$$

Let a and b be the z-components of the spin of the nuclei A and B, θ_e, φ_e and θ_ν, φ_ν the directions of motion of electron and the neutrino, and s and s_1 the spin components of the electron and the neutrino in their respective directions of motion. Starting with the nucleus A at $t = 0$ the probability amplitude at any later time t of the β-decay for given θ_e, φ_e, θ_ν, φ_ν, s, s_1, a and b will be denoted by

$$f^a_{bss_1}(\theta_e, \varphi_e, \theta_\nu, \varphi_\nu). \tag{15}$$

---

[9]See footnote 11.

Now let the electron wave function be expanded into waves $\Phi_{LJ\mathscr{P}s}$, as done before in (12) with the only difference that here $\Phi_{LJ\mathscr{P}s}$ represents a wave function with total angular momentum along the direction θ_e, φ_e (instead of along the z-axis) equal to s. The coefficients α in (12) remain unchanged. Now the $\Phi_{LJ\mathscr{P}s}$ can be further expanded into waves $\psi_{LJ\mathscr{P}s}$ with definite total angular momentum along the z-axis. The final result is

$$\sum_{LJ\mathscr{P}m} \alpha_{LJ\mathscr{P}s}(s|m)_e \psi_{LJ\mathscr{P}m} \tag{16}$$

where e represents a rotation of the coordinate axes so that the z-axis changes from the direction of motion of electron (i.e., the direction specified by θ_e, φ_e) into the laboratory z-axis. It is evident that the choice of the x and y axes perpendicular to the direction θ_e, φ_e affects only the phase of $\Phi_{LJ\mathscr{P}s}$ and would not in any way influence our final result. In (16) $(s|m)_e$ is the only factor that depends on θ_e, φ_e. A similar expansion of the neutrino wave will now be made

$$\sum_{L_1 J_1 \mathscr{P}_1 n} \beta_{L_1 J_1 \mathscr{P}_1 s_1}(s_1|n)_\nu \psi_{\nu, L_1 J_1 \mathscr{P}_1 n}. \tag{17}$$

The wave amplitude (15) is evidently given by

$$f_{bss_1}^a(\theta_e, \varphi_e, \theta_\nu, \varphi_\nu) = \sum_{\lambda\lambda_1} \sum_{mn} \alpha_{\lambda s}^* \beta_{\lambda_1 s_1}^*(s|m)_e^*(s_1|n)_\nu^* F_{b\lambda\lambda_1 mn}^a \tag{18}$$

where λ and λ_1 are abbreviations for $LJ\mathscr{P}$ and $L_1J_1\mathscr{P}_1$. We have taken the complex conjugates of the waves (16) and (17) because they represent final states. In (18) F represents the probability amplitude of the final state specified by b, $\psi_{\lambda m}$ and $\psi_{\nu,\lambda_1 n}$, the initial state being specified by a.

The probability of the β-decay is proportional to

$$\sum_{abss_1} \left| f_{bss_1}^a(\theta_e, \varphi_e, \theta_\nu, \varphi_\nu) \right|^2. \tag{19}$$

Writing

$$\sum_{ab} F_{b\lambda\lambda_1 mn}^a (F_{b\bar{\lambda}\bar{\lambda}_1 \bar{m}\bar{n}}^a)^* = G_{\Lambda m\bar{m}n\bar{n}} \tag{20}$$

and

$$\alpha_{\lambda s}^* \beta_{\lambda_1 s_1}^* \alpha_{\bar{\lambda} s} \beta_{\bar{\lambda}_1 s_1} = \Gamma_{\Lambda s s_1} \tag{21}$$

where Λ is an abbreviation for $\lambda, \lambda_1, \bar{\lambda}, \bar{\lambda}_1$, expression (19) becomes

$$\sum_{\Lambda s s_1} \Gamma_{\Lambda s s_1} \sum_{m\bar{m}n\bar{n}} G_{\Lambda m\bar{m}n\bar{n}} (s|m)_e^*(s|\bar{m})_e (s_1|n)_\nu^*(s_1|\bar{n})_\nu. \tag{22}$$

We shall show that

$$\sum_{m\bar{m}n\bar{n}} G_{\Lambda m\bar{m}n\bar{n}}(s|m)_e^*(s|\bar{m})_e(s_1|n)_\nu^*(s_1|\bar{n})_\nu \qquad (23)$$

is a polynomial of $\cos\theta$ with maximum exponent $\leq$ both $J + \bar{J}$ and $J_1 + \bar{J}_1$, θ being the angle between the directions of motion of the electron and the neutrino. But $J = L \pm \frac{1}{2}$, $J_1 = L_1 \pm \frac{1}{2}$. Hence expression (23), which represents the (cross) term in the probability of the β-decay between waves LL_1 and LL_1 is a polynomial of $\cos\theta$ with maximum exponent $\leq$ both $L + \bar{L} + 1$ and $L_1 + \bar{L}_1 + 1$.

The classification of β-decays into allowed, first forbidden etc. processes consists of an expansion in powers of $\frac{r}{\lambda_e} \sim \frac{r}{\lambda_\nu}$ ($\sim \frac{1}{10}$), λ_e, λ_ν being the wave lengths of the electron and the neutrino, and r the dimension of the nucleus. In an allowed transition only the waves $L = 0$, $L_1 = 0$ are effective for the process. Contributions from other waves are negligible because with increasing values of L the amplitude of the wave $\psi_{LJ\mathscr{P}m}$ inside the nucleus decreases as $(r/\lambda_e)^L$. In a first forbidden process $F^a_{b\lambda\lambda_1 mn}|_{L=L_1=0}$ vanishes because of selection rules and the contributing waves are the following two:[10] $L = 1$, $L_1 = 0$ and $L = 0$, $L_1 = 1$. In general for a K-th forbidden transition only waves with $L + L' \leq K$ are important.[11] This means that in the summation over Λ in (22) only $L + L_1 \leq K$, $\bar{L} + \bar{L}_1 \leq K$ terms need be retained. Hence $L_1 + \bar{L}_1 \leq 2K - (L + \bar{L})$. Thus the maximum exponent of $\cos\theta$ is $\leq$ both $L + \bar{L} + 1$ and $2K - (L + \bar{L}) + 1$, hence it is $\leq K + 1$ which proves our theorem.

It remains to be proved that the above statement about (23) is true. This we do by noticing first that F represents the probability amplitude of the final state b, $\psi_{\lambda m}$, $\psi_{\nu,\lambda_1 n}$ if the initial state is represented by a. If R is any rotation of coordinates, $\sum_a (a'|a)_R F^a_{b\lambda\lambda_1 mn}$ would give the probability amplitude of these <u>same</u> final states resulting from an initial state obtained by rotating nucleus A in state a' by R^{-1}. Thus

$$\sum_a (a'|a)_R F^a_{b\lambda\lambda_1 mn} = \sum_{b'MN} F^{a'}_{b'\lambda\lambda_1 MN}(b'|b)_R(M|m)_R(N|n)_R \qquad (24)$$

which means that $F_{\lambda\lambda_1}$ is invariant under $\mathfrak{D}^{S_A} \times \mathfrak{D}^{S_B^*} \times \mathfrak{D}^{J^*} \times \mathfrak{D}^{J_1^*}$. The definition (20) therefore shows that G_Λ is invariant under $\mathfrak{D}^{J^*} \times \mathfrak{D}^{J_1^*} \times \mathfrak{D}^{\bar{J}} \times \mathfrak{D}^{\bar{J}_1}$. I.e.

$$G_{\Lambda m\bar{m}n\bar{n}} = \sum_{M\bar{M}N\bar{N}} G_{\Lambda M\bar{M}N\bar{N}}(M|m)_R(\bar{M}|\bar{m})_R^*(N|n)_R(\bar{N}|\bar{n})_R^*.$$

---

[10] It may happen that $F^a_{b\lambda\lambda_1 mn}|_{L=L_1=0}$ is not zero but is $\sim F^a_{b\lambda\lambda_1 mn}|_{L=1,L_1=0}$. This happens in the usual interactions because of the presence of terms $\sim$ nucleon velocity. In such cases we should include the $L = 0$, $L_1 = 0$ wave. The conclusions are, however, unchanged as far as they concern only the complexity of the angular correlation.

[11] Notice that when the interaction involves derivatives of the wave function, as in the Konopinski-Uhlenbeck type of interaction, we always expand the wave function <u>before</u> taking the derivatives.

Hence
$$\sum_{m\bar{m}}(M|m)^*_R(\bar{M}|\bar{m})_R G_{\Lambda m\bar{m}n\bar{n}} = \sum_{N\bar{N}} G_{\Lambda M\bar{M}N\bar{N}}(N|n)_R(\bar{N}|\bar{n})^*_R.$$

Putting $R = \mathrm{e}$, $M = \bar{M} = s$ we see that (23) can be written

$$\sum_{N\bar{N}n\bar{n}} G_{\Lambda ssN\bar{N}}(N|n)_\mathrm{e}(\bar{N}|\bar{n})^*_\mathrm{e}(s_1|n)^*_\mathrm{v}(s_1|\bar{n})_\mathrm{v}$$

$$= \sum_{N\bar{N}n\bar{n}} G_{\Lambda ssN\bar{N}}(n|N)^*_{\mathrm{e}^{-1}}(\bar{n}|\bar{N})_{\mathrm{e}^{-1}}(s_1|n)^*_\mathrm{v}(s_1|\bar{n})_\mathrm{v} \quad (23')$$

$$= \sum_{N\bar{N}} G_{\Lambda ssN\bar{N}}(s_1|N)^*_{\mathrm{ve}^{-1}}(s_1|\bar{N})_{\mathrm{ve}^{-1}}.$$

This is evidently independent of the choice of the laboratory coordinate system. If these be so chosen that $\theta_\mathrm{e} = \varphi_\mathrm{e} = 0$ the rotation represented by e becomes the identity and (23') shows that (23) is a polynomial of $\cos\theta$ with maximum exponent $\leq J_1 + \bar{J}_1$. A similar argument shows that it is also $\leq J + \bar{J}$. This completes the proof.

If we fix our attention on one end of the spectrum where the electron momentum p is $\ll$ the neutrino momentum q, the waves that contribute most in a K-th order forbidden transition are those with $L = 0$, $L_1 \leq K$. By the theorem proved in the last section we see that the maximum exponent of $\cos\theta$ in the angular correlation is 1. This evidently applies also when $q \ll p$.

If $p \ll mc$ the spin function of the electron can be separated from the space wave function. Hence after summation over the spin directions of the electron the maximum exponent is both $\leq L + \bar{L}$ and $J_1 + \bar{J}_1 \leq 2K - (L + \bar{L}) + 1$. We have $L + \bar{L}$ instead of $J + \bar{J}$, as in all nonrelativistic cases. Thus the maximum exponent is K.

In case $p \ll q$ and $p \ll mc$, only $L = 0$ wave is effective and the angular correlation is spherically symmetrical for transitions of any order. Thus when $p \to 0$ the angular correlation becomes spherically symmetrical. On the other hand when $q \to 0$ the angular correlation becomes $1 + \alpha\cos\theta$ or 1 according as the mass of the neutrino is zero or otherwise.

Actual calculations of the angular correlation between the electron and the neutrino emitted in β-decays of different orders have been carried out by Hamilton,[12] using all the five usual types of interactions. The results, of course, conform with the theorems discussed above. Experimentally,[13] information about the angular correlation have

---

[12] D. R. Hamilton, Phys. Rev. 71, 456 (1947).
[13] J. S. Allen, Phys. Rev. 61, 692 (1942); J. C. Jacobsen, and Kofoed-Hansen, Kgl. Danske Vid. Sels., Math-fys. Medd 23, No. 12 (1945); J. S. Allen, H. R. Paneth, and A. H. Morrish, Bull. Am. Phys. Soc. 23, No. 3 (1948); C. N. Sherwin, Phys. Rev. 73, 216 (1948).

been obtained by measuring the energy spectrum of the recoil nuclei or by coincidence measurements of the electrons and the recoil nuclei. Due to the indirect nature of these experiments the results are not as yet very quantitative.

β–γ AND γ–γ CORRELATIONS

The method used in the last three sections evidently applies also to γ-rays. The rectangular components A_x, A_y, and A_z of the vector potential of the electromagnetic field is expanded into spherical harmonics. As is well known the term $L = 0$ leads to electric dipole processes, the term $L = 1$ to magnetic dipole and electric quadrupole processes etc. For each direction of propagation of the light quantum there are two possible waves with $L = 0$, corresponding to the two different polarizations. Changing the direction of propagation we obtain other waves. But altogether there are only three linearly independent waves with $L = 0$, and they transform among themselves under a rotation like a vector. Hence the angular correlation between the γ-ray and any other particle in a nuclear process is of the form $1 + \alpha \cos^2 \theta$ if the γ-ray process is of the electric dipole type. The odd power of $\cos \theta$ does not appear because the photon wave has a definite parity. This conclusion can be immediately generalized into magnetic dipole and electric quadrupole processes where the angular correlation is $1 + \alpha \cos^2 \theta + \beta \cos^4 \theta$. This holds even when both the magnetic dipole and the electric quadrupole transitions are present. Similar theorems obtain in higher multipole processes.

In general, we can study a process with any number of incoming and outgoing particles. We assume that the incoming particles are unpolarized. If one of the particles (whether incoming or outgoing), say P, has a wave length long compared to the dimension of the space–region in which it interacts with the other particles, the process can be classified according to the effective orbital angular momentum L of P. The angular correlation between P and any other particle Q in the process would then be a polynomial of $\cos \theta$ with a maximum exponent determined by L, θ being the angle between the directions of propagation of P and Q. The presence of other particles in the process does not affect the result because a summation over the directions of motion and over the spin of these "redundant" particles must always be carried out. We may say that these particles do not produce any preferential direction in space. The general results when P is a nucleon, an electron, or a photon are summarized in Table 1.

The application to the angular correlation between successive γ-rays emitted by a nucleus is straightforward. Actual calculation of this correlation for dipole–dipole, dipole–quadrupole, and quadrupole–quadrupole transitions (all electric poles) have been published.[14] They have the form

$$1 + \alpha \cos^2 \theta \quad \text{(dipole–dipole, dipole–quadrupole)}$$
$$1 + \alpha \cos^2 \theta + \beta \cos^4 \theta \quad \text{(quadrupole–quadrupole)}, \quad (25)$$

agreeing with our results. In these calculations the line width of the second γ-ray process is assumed to be large compared to the hyperfine splitting of the atom, so that the life time of the intermediate nucleus is small compared to the time required for the nuclear spin to process appreciably. Also the assumption is made that there is no magnetic dipole transition mixed with the electric quadrupole. It is evident that neither of these assumptions is necessary for the validity of our theorems, and that the angular correlation is quite generally of the form (25). It should be remarked, of course, that in case any one of these two assumptions is violated the coefficients α and β in (25) may not have the values tabulated by Hamilton.

TABLE 1

		Particle	Nuclear Particle	Electron or Neutrino	Photon
Name for different approximations	$L = 0$		S wave	Allowed	E1. dipole
	$L = 1$		P wave	First forbidden	Mag. dipole and el. quadrupole
	$L = 2$		D wave	Second forbidden	Mag. quadrupole and el. octapole
Power of $\cos \theta$			even	Even and odd	Even
Max. exponent of $\cos \theta$			$2L$	$2L + 1$	$2L + 2$

Another application is found in the problem of the angular correlation between the electron and the γ-ray emitted by a nucleus in succession. Since one of the particles is a photon only even powers of $\cos \theta$ can occur in the correlation function. Using Table 1, taking the electron to be P, we conclude that for all allowed β-transitions the correlation is spherically symmetrical. This appears at first sight very strange because e.g. for the Gamow-Teller type of interaction the matrix element involves the spin of the nucleus and one would expect that the emission of an electron in a definite direction would result in a preferential distribution of the spin orientation of the intermediate nucleus and hence would affect the angular distribution of the γ-rays. For first forbidden β-transitions the correlation is $1 + \alpha \cos^2 \theta$. Falkoff and Uhlenbeck have made actual calculations for the first forbidden-electric dipole process, using various types of β-transitions.[15] As in the

---

[14] D. R. Hamilton, Phys. Rev. 58, 122 (1940); experimental evidence has been reported by L. Brady and M. Deutsch, Phys. Rev. 72, 870 (1947).

[15] D. L. Falkoff and G. E. Uhlenbeck, Bull. Am. Phys. Soc. 22, No. 5 (1947).

γ–γ case discussed above we remark here that our conclusions hold independently of any assumption about the life time of the intermediate nucleus, and independently of the multipole nature of the γ-radiation. Also it is not necessary to neglect the term in the β-interaction that is proportional to the nucleon velocity.

REMARKS ABOUT OTHER PARTICLES

Table 1 can be extended to include mesons of spin 0 and 1. The treatment is very similar to the treatment of the electron if we use Kemmer's[16] representation of the meson wave functions. In this representation a scalar meson has a five component and a vector meson a ten component wave function. We shall assume that the rest mass is not zero. Let us take a plane wave

$$\psi = \phi e^{\frac{i}{\hbar}(\vec{p}\cdot\vec{x} - Et)} \tag{26}$$

and expand it into waves with definite orbital angular momentum L. Under a rotation the spin function ϕ is transformed by a matrix S. The total angular momentum can go as high as $L+1$. Notice that this is true for scalar mesons as well as vector mesons.[16] Thus if only orbital waves L contribute to the reaction the angular correlation between a meson and any other particle is a polynomial of $\cos\theta$ with maximum exponent $\leq 2L+2$.

If further the meson has nonrelativistic velocities v, as must actually be the case in order that the wave length of the meson may be long compared to nuclear dimensions, we can expand ϕ into a power series in v/c.

$$\phi = \phi_0 + \frac{|v|}{c}\phi_1 + \cdots \tag{27}$$

It can be readily proved that the following points are true:

(i) ϕ_0 has a definite parity and can be made independent of the direction of the velocity. The theorem proved in the section about nucleons can therefore be applied here and we see that to the order v/c the angular correlation is an even polynomial of $\cos\theta$ with maximum exponent $\leq 2L$.

(ii) ϕ_1 has a definite parity which is the opposite of that of ϕ_0. Thus the interference term between ϕ_0 and ϕ_1 gives rise to odd powers of $\cos\theta$ only and we have the result that the terms in the angular correlation to the first order of v/c is an odd polynomial of $\cos\theta$.

The author wishes to take this opportunity to thank Prof. E. Teller for invaluable discussions and advice.

---

[16] N. Kemmer, Proc. Roy. Soc. A173, 91 (1939).

ABSTRACT

Theorems concerning the general form of the angular distribution of products of nuclear reactions and disintegrations are derived. These theorems are based only on the invariance properties of the physical process under space-rotation and under inversion. The following examples are studied in detail: (i) Angular correlation between the electron and the neutrino in β-decay. (ii) Angular correlation between a β-ray and a γ-ray emitted in succession by a nucleus. (iii) Angular correlation between two γ-rays emitted in succession by a nucleus.

THE UNIVERSITY OF CHICAGO

ON THE ANGULAR DISTRIBUTION IN NUCLEAR REACTIONS
AND COINCIDENCE MEASUREMENTS

A DISSERTATION SUBMITTED TO
THE FACULTY OF THE DIVISION OF THE PHYSICAL SCIENCES
IN CANDIDACY FOR THE DEGREE OF
DOCTOR OF PHILOSOPHY

DEPARTMENT OF PHYSICS

BY
CHEN NING YANG

CHICAGO, ILLINOIS
JUNE, 1948

INTRODUCTION

In the calculation of the angular distribution in nuclear reactions and of the angular correlation in processes involving β- and γ-decay it often happens that many terms cancel out at the end of a laborious computation. The consistency of the occurrence of such cancellation leads one to suspect that some general reasons quite independent of the particular form of interaction are at work. In this paper we shall show that this is indeed the case. In fact the general form of the angular distribution in many cases can be obtained directly from the theorems derived in this paper.

For nuclear reactions between spinless particles the existence of a limitation on the complexity of the angular distribution for fixed orbital angular momentum of the incoming particles is well known. That the same result holds with the spin taken into consideration (for unpolarized incoming beam) was first pointed out by Critchfield and Teller.[1] A proof of this statement was recently given by Eisner, Sachs, and Wolfenstein.[2] We shall in this paper formulate a new proof that lends itself easily to generalization to the case in which the particles involved have relativistic velocities.

It will be shown in general that in studying the angular correlation between two particles, as long as one of them has a wave length long compared to the size of the nucleus, the process can be classified into different orders and for a process of

---

[1] C. L. Critchfield and E. Teller, Phys. Rev. **60**, 10 (1941).

[2] E. Eisner and R. G. Sachs, Phys. Rev. **72**, 680 (1947); L. Wolfenstein and R. G. Sachs, **73**, 528 (1948).

2

given order the general form of the angular correlation is essentially known. In case both of the particles have long wave lengths particularly simple conclusions may be reached, as in the case of β-neutrino correlation in β-decay.

Experimentally the angular correlations β-neutrino and γ-γ have been studied by many authors. Various calculations of these correlations based on different kinds of interactions have also been made. These will be separately discussed in the different sections.

NUCLEAR REACTION

Consider the following reaction
$$A + P \longrightarrow B + Q \tag{1}$$
and suppose both the target nucleus A and the bombarding beam of particles P are unpolarized. The complexity of the angular distribution of the outgoing particles is limited by the following theorem: <u>If only incoming waves of orbital angular momentum L contribute appreciably to the reaction, the angular distribution of the outgoing particles in the centre of mass system is an even polynomial of $\cos\theta$ with maximum exponent not higher than 2L.</u> Here θ is the angle between the incoming and the outgoing particles in the centre of mass system of reference.

To prove this let us consider the collision between two particles A and P with definite (a and p) components of spin along the z-axis, and definite total and z-component relative orbital angular momenta L and m. (We use the centre of mass system throughout.) The incoming wave function is, at large distances r_{AP} between A and P:

$$\frac{1}{r_{AP}} \sin(k_{AP} r_{AP} - \tfrac{1}{2} L\pi)\, \psi_A^a\, \psi_P^p\, Y_{Lm}(\theta_P, \phi_P) \tag{2}$$

where ψ_A^a, ψ_P^p are normalized internal wave functions of particles A and P; θ_P, ϕ_P describe the direction of motion of the particle P; and $Y_{Lm}(\theta,\phi)$ is the normalized spherical harmonic of order Lm.

The asymtotic behavior of the wave function at large values of r_{BQ} is of the form

$$\frac{1}{r_{BQ}} \exp(ik_{BQ} r_{BQ}) \sum_{bq} \psi_B^b\, \psi_Q^q\, f_{bq}^{apm}(\theta_Q, \phi_Q). \tag{3}$$

3

4

In reaction (1) if we choose as the z-axis the direction of motion of particle P, it is clear that when the incoming wave is expanded into partial waves with definite total and z-component orbital angular momenta L and m only terms with m = 0 occur. Under the assumption stated in the theorem we can neglect all terms except the spherical harmonic Y_{L0}. The differential cross-section of reaction (1) is therefore

$$d\sigma = (\text{constant}) \, d\Omega_Q \sum_{bq} |f^{apo}_{bq}(\theta_Q, \phi_Q)|^2. \tag{4}$$

For unpolarized incoming particles we get

$$d\sigma = (\text{constant}) \, d\Omega_Q \sum_{apbq} |f^{apo}_{bq}(\theta, \phi)|^2. \tag{5}$$

The requirement of invariance under rotation will now be introduced. Consider a new coordinate system (primed system) obtained from the old by a rotation of the coordinate axis. Let $(m'|m)^{(L)}$ be the matrix element[3] of the irreducible representation D^L of the three dimensional rotation group. We have

$$Y_{Lm'}(\theta', \phi') = \sum_m (m'|m)^{(L)} Y_{Lm}(\theta, \phi),$$
$$\psi_A'^{a'} = \sum_a (a'|a)^{(S_A)} \psi_A^a, \tag{6}$$

where S_A = spin of particle A, which may be an integer or a half odd integer. By $\psi_A'^{a'}$ is meant the function $\psi_A^{a'}$ of the <u>primed</u> (internal) coordinates. The proof of our theorem consists in showing that (i) the superposition principle requires that f be transformed according to $D^{S_A} \times D^{S_P} \times D^L \times D^{S_B^*} \times D^{S_Q^*}$ and (ii) the expression

$$I^{mm'}(\theta, \phi) = \sum_{apbq} \left[f^{apm}_{bq}(\theta, \phi)\right]^* f^{apm'}_{bq}(\theta, \phi) \tag{7}$$

transforms according to $D^{L^*} \times D^L = D^{2L} + D^{2L-2} + \cdots$. No non-vanishing linear combinations of (7) transform according to $D^{2L-1}, D^{2L-3} \cdots$ or D^l.

---

[3] E. Wigner, Gruppentheorie und ihre Anwendung auf die Quantenmechanik der Atomspektren (1931), 180.

5

(i) Consider the following incoming wave

$$\frac{1}{r_{AP}} \sin(k_{AP}r_{AP} - \tfrac{1}{2}L\pi) \psi_A^{'a'} \psi_P^{'p'} Y_{Lm}(\theta_P', \phi_P'). \qquad (2')$$

To an observer in the primed coordinate system this has exactly the same form as (2). Hence the outgoing wave must be

$$\frac{1}{r_{BQ}} \exp(ik_{BQ}r_{BQ}) \sum_{b'q'} \psi_B^{'b'} \psi_Q^{'q'} f_{b'q'}^{a'p'm'}(\theta_Q', \phi_Q'). \qquad (3')$$

Notice that we use the <u>same</u> f instead of an f', because there is <u>no physically observable distinction</u> between the two coordinate systems. Using (5) one can express (2') as a superposition of waves (2)

$$\sum_{apm}(a'|a)(p'|p)(m'|m)\left[\frac{1}{r_{AP}} \sin(k_{AP}r_{AP} - \tfrac{1}{2}L\pi) \psi_A^a \psi_P^p Y_{Lm}(\theta_P, \phi_P)\right].$$

Here we have omitted the superscripts S_A, S_P, L from $(a'|a)^{(S_A)}$, $(p'|p)^{(S_P)}$, $(m'|m)^{(L)}$ for simplicity. The outgoing wave must therefore be a corresponding superposition of waves (3) with the same coefficients:

$$\sum_{apm}(a'|a)(p'|p)(m'|m)\left[\frac{1}{r_{BQ}} \exp(ik_{BQ}r_{BQ}) \sum_{bq} \psi_B^b \psi_Q^q f_{bq}^{apm}(\theta_Q, \phi_Q)\right].$$

Equating this to (3') and using (6) to express $\psi_B^{'b'}$, $\psi_Q^{'q'}$ in terms of ψ_B^b, ψ_Q^q we get finally by identifying the coefficient of $\psi_B^b \psi_Q^q$:

$$\sum_{apm}(a'|a)(p'|p)(m'|m) f_{bq}^{apm}(\theta,\phi) = \sum_{b'q'}(b'|b)(q'|q) f_{b'q'}^{a'p'm'}(\theta',\phi').$$

This reduces to the following form

$$f_{b'q'}^{a'p'm'}(\theta',\phi') = \sum_{apmbq}(a'|a)(p'|p)(m'|m)(b'|b)^*(q'|q)^* f_{bq}^{apm}(\theta,\phi) \qquad (8)$$

through the orthogonality relations

$$\sum_m (m'|m)(m''|m) = \delta_{m'm''}. \qquad (9)$$

Equation (8) expresses the transformation property of f.

(ii) To obtain the transformation property of expression (4) we investigate the behavior of expression (7) under rotation. By (8) and (9)

6

$$I^{m''m'}(\theta',\varphi') = \sum_{mm'''}\sum_{abpq} (m''|m''')^*(m'|m) [f^{apm'''}_{bq}(\theta,\phi)]^* f^{apm}_{bq}(\theta,\phi)$$

$$= \sum_{mm'''}(m''|m''')^*(m'|m) I^{m'''m}(\theta,\varphi). \qquad (10)$$

Now the differential cross-section is proportional to I^{oo}. If we put in this expression $m'' = m' = o$ and take the rotation from the unprimed to the primed coordinate system to be a rotation around the z-axis by an angle ξ we have $\theta' = \theta$ and $\varphi' = \varphi + \xi$. Since then $(o|m) = \delta_{mo}$ it is evident from (10) that

$$I^{oo}(\theta,\varphi+\xi) = I^{oo}(\theta,\varphi)$$

showing that I^{oo} is independent of φ. To study its dependence on θ we put in (10) $m' = m'' = o$. It is well known[4] that if $\theta = \varphi = o$

$$(o|m) = Y_{L,-m}(\theta',\varphi').$$

Hence (10) becomes

$$I^{oo}(\theta',\varphi') = \sum_{m\,m'''} Y^*_{L,-m'''}(\theta',\varphi') Y_{L,-m}(\theta',\varphi') I^{m'''m}(o,o).$$

On application of the reduction theorem[4] of products of spherical harmonics this leads directly to our theorem.

If instead of a rotation we had chosen an inversion of the coordinates it is evident that (8) would become

$$f^{apm}_{bq}(\pi-\theta,\pi+\varphi) = \mathcal{P}_A\mathcal{P}_P\mathcal{P}_B\mathcal{P}_Q(-1)^L f^{apm}_{bq}(\theta,\varphi) \qquad (11)$$

where $\mathcal{P}_A$, $\mathcal{P}_P$ etc. are the intrinsic parities of the nuclei. This shows that

$$|f(\pi-\theta,\pi+\varphi)|^2 = |f(\theta,\varphi)|^2$$

and it follows that the angular dependence must be an even function of $\cos\theta$, a fact that is already established by (10). Equation (11) further shows that any odd power of $\cos\theta$ in the angular dependence must come from an interference term between orbital wave functions of opposite parity.

---

[4] H. Bethe, <u>Handbuch der Physik</u>, Band 24/1, Kap. 3, Section 65.

7

The symmetry requirements of the wave function under interchanges of the nucleons do not[5] in general lead to any new conclusions about the properties of f. However in the special case in which the two incoming particles or the two outgoing particles are identical more detailed consideration is necessary. An example of such a case is the reaction

$$Li^7 + H^1 \rightarrow He^4 + He^4.$$

Since the outgoing particles are spinless and satisfy Bose-Einstein statistics and since Li^7 has an odd parity, the value of L must be odd in order to have a balance of parity. This means that $f = 0$ unless L is odd. At low energies, therefore, the effective orbital angular momentum of the incoming particles is 1.

Another example is the $D^2 + D^2$ reaction:

$$D^2 + D^2 \rightarrow N^1 + He^3,$$
$$D^2 + D^2 \rightarrow H^1 + H^3.$$

This reaction has recently been considered theoretically by Konopinski and Teller.[6] Due to the symmetry nature of the deuterons it is no longer convenient to specify the spin of the two incoming particles separately. Instead we should group the nine possible incoming states into a quintet, a triplet, and a singlet. The space wave functions for the quintet and the singlet states are symmetrical with respect to the exchange of the two deuterons and those for the triplet states are antisymmetrical. Strictly speaking the proof of our theorem does not apply to such a case

---

[5] To understand this it is best to introduce the idea of channels in the configuration space, which was first discussed by G. Breit, Phys. Rev. **58**, 1068 (1940); J. A. Wheeler, Phys. Rev. **52**, 1107 (1937). An interchange of the nucleons in general results in an interchange of the channels, except for the case when either A and P or B and Q are identical.

[6] E. J. Konopinski and E. Teller, Phys. Rev. **73**, 822 (1948).

8

where the space wave function depends on the orientation of the spins of the particles. But since all the states in the same multiplet have the same *a priori* probability it is evident that the difference of the space wave function for the different multiplets does not affect the validity of our theorem.

The Coulomb field affects the waves of different orbital angular momenta in such a way as to favor those with higher angular momenta at low energies.[7] This accounts for the reason why at bombarding energies as low as 20 kev. the angular distribution in the D + D reaction is not symmetrical spherically.[6,8] We shall not go into this point in any further details here.

We conclude this section by stating a variation of the theorem proved above: When contributions from incoming waves with orbital angular momenta L are neglected the angular distribution in reaction (1) in the centre of mass system is a polynomial of $\cos\theta$ with maximum exponent not higher than 2L. This holds even if the contributing compound nuclear states have angular momenta >L.

It will be noticed that when both even and odd values of the orbital angular momenta in the incoming beam are effective in producing the reaction the angular distribution contains odd powers of $\cos\theta$. This, however, will not happen when either (i) the reaction goes through a single compound nuclear state (e.g. near a strong resonance level); or (ii) symmetry requirements exclude even (or odd) L values as in the $Li^7 + H^1 \rightarrow He^4 + He^4$ reaction discussed above.

---

[7] H. Bethe and E. J. Konopinski, Phys. Rev. **54**, 130 (1938).

[8] E. Bretscher, A. P. French, and F. G. P. Seidl, Phys. Rev. **73**, 815 (1948).

RELATIVISTIC CASE

We shall in this section generalize the result of the last section to the case when the particle P is an electron and has relativistic velocities. (The nuclei A, B, and Q are still supposed to be nonrelativistic.) No such process has been experimentally realized. We shall, however, discuss it to illustrate our method. It will be proved that <u>if only partial waves of orbital angular momentum L in the electron wave function contribute to the reaction the angular correlation is a polynomial of $\cos\theta$ with maximum exponent not higher than $2L + 1$</u>.

Instead of the stationary picture used in the last section we shall here use a nonstationary description of the process. The electron wave function at time $t = 0$ is a product of a spin wave function with four components and a space wave function e^{ikz}. The spin of the electron along the z-axis is a constant of motion and is denoted by $p(=\pm\frac{1}{2})$. If we expand the space wave function into partial waves of definite orbital angular momenta L, the first term (L = 0) would give rise to allowed transitions, the second term (L = 1) first forbidden transitions etc. To study the angular correlation arising from the contribution of the partial wave of orbital angular momentum L we need to decompose it again into* normalized waves $\Psi_{LJ\wp m}$ of definite L, J (total angular momentum of the electron), $\wp$ (parity), and m (z-component of $\vec{J}$). The advantage of using these Ψ's is that they have simple transformation properties under rotation. The possible values of J are $L \pm \frac{1}{2}$. Under the assumption that we are only considering the contribution from a definite L value the wave function at $t = 0$ can be replaced by

10

$$\sum_{\mathcal{P}=\pm 1}\sum_{J=L\pm\frac{1}{2}} \alpha_{LJ\mathcal{P}p}\, \psi_{LJ\mathcal{P}p}. \tag{12}$$

We have put $m = p$ because the z-component of the orbital angular momentum is zero. $\mathcal{P}$ ranges over 1 and -1.*

Let us now first study the reaction arising from the electron wave $\psi_{LJ\mathcal{P}m}$. Starting at $t = 0$ with $\psi_{LJ\mathcal{P}m}$ and nucleus A with a definite value a for the z-component of spin, we shall denote by $f^{LJ\mathcal{P}am}_{bq}(\theta_Q, \varphi_Q)$ the <u>probability amplitude</u> at any later time $t > 0$ for that outgoing state in which the z-component of spin of the particles B and Q are b and q, and in which the momentum of Q is in the direction θ_Q, φ_Q. The absolute value of the outgoing momentum (which is not fixed because the energy is not necessarily conserved when t is small) should also enter the function f as an independent variable, but has been omitted for simplicity of writing.

Now the probability amplitudes are additive when we superpose states. Since under a rotation the different waves $\psi_{LJ\mathcal{P}m}$ with the same $L, J, \mathcal{P}$ values combine linearly, the argument which led to (8) in the last section would now lead to

$$f^{LJ\mathcal{P}a'm'}_{b'q'}(\theta', \varphi') = \sum_{ambq} (a'|a)(b'|b)^*(m'|m)(q'|q)^* f^{LJ\mathcal{P}am}_{bq}(\theta,\varphi). \tag{13}$$

Returning now to the wave (12) at $t = 0$ we see that the differential cross-section is proportional to

$$d\Omega_Q \left| \sum_{\mathcal{P}J} \alpha_{LJ\mathcal{P}p}\, f^{LJ\mathcal{P}ap}_{bq}(\theta,\varphi) \right|^2.$$

This will have to be summed over a, b, p, and q. Since the coefficients in (12') are independent of a, b, and q the final expression is

$$d\Omega_Q \sum_{J\mathcal{P}J'\mathcal{P}'p} \alpha_{LJ\mathcal{P}p}\, \alpha^*_{LJ'\mathcal{P}'p}\left[\sum_{abq}\left\{f^{LJ'\mathcal{P}'ap}_{bq}(\theta,\varphi)\right\}^* f^{LJ\mathcal{P}ap}_{bq}(\theta,\varphi)\right]. \tag{14}$$

By (13) the individual terms under the summation sign $\sum_{abq}$

---

*The parity can be either 1 or -1 for any given L, J and m. However, for slow electrons the amplitude of waves with $\mathcal{P} = -(-1)^L$ is very small. Cf. the end of this section.

11

transform under a rotation according to $D^{S_a}_x D^{S_a*}_x D^{S_b}_x D^{S_b*}_x D^{S_q}_x D^{S_q}_x D^{J}_x D^{J'*}_x$. But after the summation over a, b, and q is carried out, the sum transforms more simply according to $D^J_x D^{J'*} = D^{J+J'} + D^{J+J'-1} + \cdots$. This means that the expression in the square bracket in (14) is a sum of spherical harmonics of order Lm with $L \leq J + J'$. But both J and J' are $\leq L + \frac{1}{2}$. The theorem stated at the beginning of this section follows immediately.

If we introduce the requirement of invariance under inversion (14) shows that those terms with $\mathcal{P}'\mathcal{P}$ = even give rise to angular correlation functions that are even under the transformation $\theta \to \pi - \theta$; and those with $\mathcal{P}'\mathcal{P}$ = odd give rise to odd angular correlation functions. A consequence of this is the following. If the velocity v of the electron is small compared to the velocity of light c, and if the spin wave function of the electron is expanded in powers of $\frac{v}{c}$, the first term, i.e. the term that does not vanish as $v \to 0$, is invariant under an inversion. This term would therefore give rise to terms with $\mathcal{P} = (-1)^J$. The opposite parity first appears in the next term of the expansion and is proportional to $\frac{v}{c}$. Hence those terms in (14) with $\mathcal{P}'\mathcal{P}$ = odd contain a factor $\frac{v}{c}$. Thus <u>the odd powers of $\cos\theta$ in the angular correlation have coefficients smaller than the even powers by a factor of $\frac{v}{c}$</u>.

β-NEUTRINO CORRELATION

In β-decay we have the particularly simple situation in which both the electron and the neutrino have wave lengths long compared to the dimension of the nucleus. The argument of the last section can now be applied to both these particles and we can prove that <u>the angular correlation between the electron and the neutrino emitted in a β-decay is a polynomial of $\cos\theta$ up to a maximum exponent $K + 1$, where[9] $K = 0$ for allowed transitions, $K = 1$ for first forbidden transitions etc.</u>

The idea of the proof is that for first forbidden transitions one has either $L = 1$ for the electron and $L_1 = 0$ for the neutrino or $L = 0$ for the electron and $L_1 = 1$ for the neutrino. The waves $L = 1$ and $L_1 = 1$ occur together only in second forbidden processes. Now the intensity produced by the $L = 1$, $L_1 = 0$ waves has an angular correlation function that goes up to $\cos$ to the first power, according to the theorem of the last section. Similarly fixing our attention on the neutrino wave function we can draw the same conclusion about the $L = 0$, $L_1 = 1$ waves. The interference term of the $L = 1$, $L_1 = 0$ waves with the $L = 0$, $L_1 = 1$ waves, however, gives an angular distribution that contains $\cos^2\theta$, which is the highest power of $\cos\theta$ possible for this case.

The proof is as follows. Consider the β-decay

$$A \rightarrow B + e + \nu .$$

Let a and b be the z-components of the spin of the nuclei A and B, θ_e, φ_e and θ_ν, φ_ν the directions of motion of the electron and

---

[9]See footnote 11.

13

the neutrino, and s and s_1 the spin components of the electron and the neutrino in their respective directions of motion. Starting with the nucleus A at t = 0 the probability amplitude at any later time t of the β-decay for given θ_e, φ_e, θ_ν, φ_ν, s, s_1, a and b will be denoted by

$$f^a_{bss_1}(\theta_e, \varphi_e, \theta_\nu, \varphi_\nu). \tag{15}$$

Now let the electron wave function be expanded into waves $\Phi_{LJ\varphi S}$, as done before in (12) with the only difference that here $\Phi_{LJ\varphi S}$ represents a wave function with total angular momentum along the direction θ_e, φ_e (instead of along the z-axis) equal to s. The coefficients α in (12) remain unchanged. Now the $\Phi_{LJ\varphi S}$ can be further expanded into waves $\Psi_{LJ\varphi S}$ with definite total angular momentum along the z-axis. The final result is

$$\sum_{LJ\varphi m} \alpha_{LJ\varphi S}(s|m)_e \Psi_{LJ\varphi m} \tag{16}$$

where e represents a rotation of the coordinate axes so that the z-axis changes from the direction of motion of the electron (i.e. the direction specified by θ_e, φ_e) into the laboratory z-axis. It is evident that the choice of the x and y axes perpendicular to the direction θ_e, φ_e affects only the phase of $\Phi_{LJ\varphi S}$ and would not in any way influence our final result. In (16) $(s|m)_e$ is the only factor that depends on θ_e, φ_e. A similar expansion of the neutrino wave will now be made

$$\sum_{L,J,\varphi_1,n} \beta_{L,J,\varphi_1,s_1}(s_1|n)_\nu \Psi_{\nu,L,J,\varphi_1,n}. \tag{17}$$

The wave amplitude (15) is evidently given by

$$f^a_{bss_1}(\theta_e, \varphi_e, \theta_\nu, \varphi_\nu) = \sum_{\lambda\lambda_1}\sum_{mn} \alpha^*_{\lambda s} \beta^*_{\lambda_1 s_1}(s|m)^*_e (s_1|n)^*_\nu F^a_{b\lambda\lambda_1 mn} \tag{18}$$

where λ and λ_1 are abbreviations for $LJ\varphi$ and L,J,φ_1. We have taken the complex conjugates of the waves (16) and (17) because they represent final states. In (18) F represents the probability amplitude of the final state specified by b, $\Psi_{\lambda m}$ and $\Psi_{\nu,\lambda_1 n}$, the initial state being specified by a.

14

The probability of the β-decay is proportional to

$$\sum_{abss_1} | f^a_{bss_1}(\theta_e, \varphi_e, \theta_\nu, \varphi_\nu) |^2 . \tag{19}$$

Writing

$$\sum_{ab} F^a_{b\lambda\lambda_1 mn} (F^a_{b\bar{\lambda}\bar{\lambda}_1 \bar{m}\bar{n}})^* = G_{\Lambda m\bar{m} n\bar{n}} \tag{20}$$

and

$$\alpha^*_{\lambda s} \beta^*_{\lambda_1 s_1} \alpha_{\bar{\lambda} s} \beta_{\bar{\lambda}_1 s_1} = \Gamma_{\Lambda s s_1} \tag{21}$$

where Λ is an abbreviation for $\lambda, \lambda_1, \bar{\lambda}, \bar{\lambda}_1$, expression (19) becomes

$$\sum_{\Lambda s s_1} \Gamma_{\Lambda s s_1} \sum_{m\bar{m}n\bar{n}} G_{\Lambda m\bar{m}n\bar{n}} (s|m)^*_e (s|\bar{m})_e (s_1|n)^*_\nu (s_1|\bar{n})_\nu . \tag{22}$$

We shall show that

$$\sum_{m\bar{m}n\bar{n}} G_{\Lambda m\bar{m}n\bar{n}} (s|m)^*_e (s|\bar{m})_e (s_1|n)^*_\nu (s_1|\bar{n})_\nu \tag{23}$$

is a polynomial of $\cos\theta$ with maximum exponent $\leq$ both $J + \bar{J}$ and $J_1 + \bar{J}_1$, θ being the angle between the directions of motion of the electron and the neutrino. But $J = L \pm \frac{1}{2}$, $J_1 = L_1 \pm \frac{1}{2}$. Hence expression (23), which represents the (cross) term in the probability of the β-decay between waves LL_1 and $\bar{L}\bar{L}_1$ is a polynomial of $\cos\theta$ with maximum exponent $\leq$ both $L + \bar{L} + 1$ and $L_1 + \bar{L}_1 + 1$.

The classification of β-decays into allowed, first forbidden etc. processes consists of an expansion in powers of $\frac{r}{\lambda_e} \sim \frac{r}{\lambda_\nu} (\sim \frac{1}{10})$, λ_e, λ_ν being the wave lengths of the electron and the neutrino, and r the dimension of the nucleus. In an allowed transition only the waves $L = 0$, $L_1 = 0$ are effective for the process. Contributions from other waves are negligible because with increasing values of L the amplitude of the wave $\psi_{LJ\sigma m}$ inside the nucleus decreases as $\left(\frac{r}{\lambda_e}\right)^L$. In a first forbidden process $F^a_{b\lambda\lambda_1 mn}\big|_{L=L_1=0}$ vanishes because of selection rules and the contributing waves are the following two:[10] $L = 1$, $L_1 = 0$ and $L = 0$,

---

[10] It may happen that $F^a_{b\lambda\lambda_1 mn}\big|_{L=L_1=0}$ is not zero but is $\sim F^a_{b\lambda\lambda_1 mn}\big|_{L=1, L_1=0}$. This happens in the usual interactions because

15

$L_1 = 1$. In general for a K-th forbidden transition only waves with $L + L' \leq K$ are important.[11] This means that in the summation over Λ in (22) only $L + L_1 \leq K$, $\bar{L} + \bar{L}_1 \leq K$ terms need be retained. Hence $L_1 + \bar{L}_1 \leq 2K-(L + \bar{L})$. Thus the maximum exponent of $\cos\theta$ is $\leq$ both $L + \bar{L} + 1$ and $2K-(L + \bar{L}) + 1$, hence it is $\leq K + 1$ which proves our theorem.

It remains to be proved that the above statement about (23) is true. This we do by noticing first that F represents the probability amplitude of the final state b, $\psi_{\lambda m}$, $\psi_{\nu, \lambda_1 n}$ if the initial state is represented by a. If R is any rotation of coordinates, $\sum_a (a'|a)_R F^a_{b\lambda\lambda_1,mn}$ would give the probability amplitude of these <u>same</u> final states resulting from an initial state obtained by rotating nucleus A in state a' by R^{-1}. Thus

$$\sum_a (a'|a)_R F^a_{b\lambda\lambda_1,mn} = \sum_{b'MN} F^{a'}_{b'\lambda\lambda_1,MN} (b'|b)_R (M|m)_R (N|n)_R \quad (24)$$

which means that $F_{\lambda\lambda_1}$ is invariant under $\mathcal{D}^{S_A} \times \mathcal{D}^{S_B*} \times \mathcal{D}^{J*} \times \mathcal{D}^{J_1*}$. The definition (20) therefore shows that G_Λ is invariant under $\mathcal{D}^{J*} \times \mathcal{D}^{\bar{J}*} \times \mathcal{D}^{J} \times \mathcal{D}^{\bar{J}_1}$. I.e.

$$G_{\Lambda m\bar{m}n\bar{n}} = \sum_{M\bar{M}N\bar{N}} G_{\Lambda M\bar{M}N\bar{N}} (M|m)_R (\bar{M}|\bar{m})^*_R (N|n)_R (\bar{N}|\bar{n})^*_R .$$

Hence

$$\sum_{m\bar{m}} (M|m)^*_R (\bar{M}|\bar{m})_R G_{\Lambda m\bar{m}n\bar{n}} = \sum_{N\bar{N}} G_{\Lambda M\bar{M}N\bar{N}} (N|n)_R (\bar{N}|\bar{n})^*_R .$$

Putting $R = e$, $M = \bar{M} = s$ we see that (23) can be written

$$\sum_{N\bar{N}n\bar{n}} G_{\Lambda ss N\bar{N}} (N|n)_e (\bar{N}|\bar{n})^*_e (s_1|n)^*_\nu (s_1|\bar{n})_\nu$$
$$= \sum_{N\bar{N}n\bar{n}} G_{\Lambda ss N\bar{N}} (n|N)^*_{e^{-1}} (\bar{n}|\bar{N})_{e^{-1}} (s_1|n)^*_\nu (s_1|\bar{n})_\nu$$
$$= \sum_{N\bar{N}} G_{\Lambda ss N\bar{N}} (s_1|N)^*_{\nu e^{-1}} (s_1|\bar{N})_{\nu e^{-1}} . \qquad (23')$$

---

of the presence of terms $\sim$ nucleon velocity. In such cases we should include the $L = 0$, $L_1 = 1$ wave. The conclusions are, however, unchanged as far as they concern only the complexity of the angular correlation.

[11] Notice that when the interaction involves derivatives of the wave function, as in the Konopinski-Uhlenbeck type of interaction, we always expand the wave function <u>before</u> taking the derivatives.

16

This is evidently independent of the choice of the laboratory coordinate system. If these be so chosen that $\theta_e \doteq \varphi_e = 0$ the rotation represented by e becomes the identity and (23') shows that (23) is a polynomial of $\cos\theta$ with maximum exponent $\leq J_1 + \bar{J}_1$. A similar argument shows that it is also $\leq J + \bar{J}$. This completes the proof.

If we fix our attention on one end of the spectrum where the electron momentum p is $\ll$ the neutrino momentum q, the waves that contribute most in a K-th order forbidden transition are those with $L = 0$, $L_1 \leq K$. By the theorem proved in the last section we see that <u>the maximum exponent of $\cos\theta$ in the angular correlation is 1.</u> This evidently applies also when $q \ll p$.

If $p \ll mc$ the spin function of the electron can be separated from the space wave function. Hence after summation over the spin directions of the electron the maximum exponent is both $\leq L + \bar{L}$ and $J_1 + \bar{J}_1 \leq 2K-(L + \bar{L}) + 1$. We have $L + \bar{L}$ instead of $J + \bar{J}$, as in all nonrelativistic cases. Thus <u>the maximum exponent is K.</u>

<u>In case $p \ll q$ and $p \ll mc$, only $L = 0$ wave is effective and the angular correlation is spherically symmetrical for transitions of any order.</u> Thus when $p \to 0$ the angular correlation becomes spherically symmetrical. On the other hand when $q \to 0$ the angular correlation becomes $1 + \alpha\cos\theta$ or 1 according as the mass of the neutrino is zero or otherwise.

Actual calculations of the angular correlation between the electron and the neutrino emitted in β-decays of different orders have been carried out by Hamilton,[12] using all the five usual types of interactions. The results, of course, conform

---

[12] R. R. Hamilton, Phys. Rev. **71**, 456 (1947).

17

with the theorems discussed above. Experimentally[13] information about the angular correlation have been obtained by measuring the energy spectrum of the recoil nuclei or by coincidence measurements of the electrons and the recoil nuclei. Due to the indirect nature of these experiments the results are not as yet very quantitative.

---

[13] J. S. Allen, Phys. Rev. 61, 692 (1942); J. C. Jacobsen and Kofoed-Hansen, Kgl. Danske Vid. Sels., Math-fys. Medd. 23, No. 12 (1945); J. S. Allen, H. R. Paneth, and A. H. Morrish, Bull. Am. Phys. Soc. 23, No. 3 (1948); C. N. Sherwin, Phys. Rev. 73, 216 (1948).

β-γ AND γ-γ CORRELATIONS

The method used in the last three sections evidently applies also to γ-rays. The rectangular components A_x, A_y, and A_z of the vector potential of the electromagnetic field is expanded into spherical harmonics. As is well known the term $L = 0$ leads to electric dipole processes, the term $L = 1$ to magnetic dipole and electric quadrupole processes etc. For each direction of propagation of the light quantum there are two possible waves with $L = 0$, corresponding to the two different polarizations. Changing the direction of propagation we obtain other waves. But altogether there are only three linearly independent waves with $L = 0$, and they transform among themselves under a rotation like a vector. Hence the angular correlation between the γ-ray and any other particle in a nuclear process is of the form $1 + \alpha \cos^2\theta$ if the γ-ray process is of the electric dipole type. The odd power of $\cos\theta$ does not appear because the photon wave has a definite parity. This conclusion can be immediately generalized into magnetic dipole and electric quadrupole processes where the angular correlation is $1 + \alpha\cos^2\theta + \beta\cos^4\theta$. This holds even when both the magnetic dipole and the electric quadrupole transitions are present. Similar theorems obtain in higher multipole processes.

In general we can study a process with any number of incoming and outgoing particles. We assume that the incoming particles are unpolarized. If one of the particles (whether incoming or outgoing), say P, has a wave length long compared to the dimension of the space-region in which it interacts with the other particles, the process can be classified according to the

19

effective orbital angular momentum L of P. The angular correlation between P and any other particle Q in the process would then be a polynomial of $\cos\theta$ with a maximum exponent determined by L, θ being the angle between the directions of propagation of P and Q. The presence of other particles in the process does not affect the result because a summation over the directions of motion and over the spin of these "redundant" particles must always be carried out. We may say that these particles do not produce any preferential direction in space. The general results when P is a nucleon, an electron, or a photon are summarized in Table 1.

The application to the angular correlation between successive γ-rays emitted by a nucleus is straightforward. Actual calculation of this correlation for dipole-dipole, dipole-quadrupole, and quadrupole-quadrupole transitions (all electric poles) have been published.[14] They have the form

$$1 + \alpha\cos^2\theta \quad \text{(dipole-dipole, dipole-quadrupole)}$$
$$1 + \alpha\cos^2 + \beta\cos^4\theta \quad \text{(quadrupole-quadrupole)}, \quad (25)$$

agreeing with our results. In these calculations the line width of the second γ-ray process is assumed to be large compared to the hyperfine splitting of the atom, so that the life time of the intermediate nucleus is small compared to the time required for the nuclear spin to precess appreciably. Also the assumption is made that there is no magnetic dipole transition mixed with the electric quadrupole. It is evident that neither of these assumptions is necessary for the validity of our theorems, and that the angular correlation is quite generally of the form (25). It should be remarked, of course, that in case any one of these two

---

[14] D. R. Hamilton, Phys. Rev. 58, 122 (1940); experimental evidence has been reported by L. Brady and M. Deutsch, Phys. Rev. 72, 870 (1947).

20

assumptions is violated the coefficients α and β in (25) may not have the values tabulated by Hamilton.

TABLE 1

Particle		Nuclear Particle	Electron or Neutrino	Photon
Name for different approximations	L = 0	S wave	Allowed	El. dipole
	L = 1	P wave	First forbidden	Mag. dipole and el. quadrupole
	L = 2	D wave	Second forbidden	Mag. quadrupole and el. octapole
Power of $\cos\theta$		even	Even and odd	Even
Max. exponent of $\cos\theta$		2L	2L + 1	2L + 2

Another application is found in the problem of the angular correlation between the electron and the γ-ray emitted by a nucleus in succession. Since one of the particles is a photon only even powers of $\cos\theta$ can occur in the correlation function. Using Table 1, taking the electron to be P, we conclude that <u>for all allowed β-transitions the correlation is spherically symmetrical.</u> This appears at first sight very strange because e.g. for the Gamow-Teller type of interaction the matrix element involves the spin of the nucleus and one would expect that the emmission of an electron in a definite direction would result in a preferential distribution of the spin orientation of the intermediate nucleus and hence would affect the angular distribution of the γ-rays. <u>For first forbidden β-transitions the correlation is $1 + \alpha \cos^2\theta$.</u> Falkoff and Uhlenbeck have made actual calculations for the first forbidden-electric dipole process, using various types of β-interactions.[15] As in the γ-γ case discussed above we remark here that our conclusions hold independently of

---

[15] D. L. Falkoff and G. E. Uhlenbeck, Bull. Am. Phys. Soc. 22, No. 5 (1947).

21

any assumption about the life time of the intermediate nucleus, and independently of the multipole nature of the γ-radiation. Also it is not necessary to neglect the term in the β-interaction that is proportional to the nucleon velocity.

REMARKS ABOUT OTHER PARTICLES

Table 1 can be extended to include mesons of spin 0 and 1. The treatment is very similar to the treatment of the electron if we use Kemmer's[16] representation of the meson wave functions. In this representation a scalar meson has a five component and a vector meson a ten component wave function. We shall assume that the rest mass is not zero. Let us take a plane wave

$$\psi = \phi e^{\frac{i}{\hbar}(\vec{p}\cdot\vec{x} - Et)} \tag{26}$$

and expand it into waves with definite orbital angular momentum L. Under a rotation the spin function ϕ is transformed by a matrix S. The total angular momentum can go as high as $L + 1$. Notice that this is true for scalar mesons as well as vector mesons.[16] Thus if only orbital waves L contribute to the reaction the angular correlation between a meson and any other particle is a polynomial of $\cos\theta$ with maximum exponent $\leq 2L + 2$.

If further the meson has nonrelativistic velocities v, as must actually be the case in order that the wave length of the meson may be long compared to nuclear dimensions, we can expand ϕ into a power series in $\frac{v}{c}$.

$$\phi = \phi_0 + \frac{|v|}{c}\phi_1 + \cdots . \tag{27}$$

It can be readily proved that the following points are true:

(i) ϕ_0 has a definite parity and can be made independent of the direction of the velocity. The theorem proved in the section about nucleons can therefore be applied here and we see that to the order $\frac{v}{c}$ the angular correlation is an even

---

[16] N. Kemmer, Proc. Roy. Soc. **A173**, 91 (1939).

23

polynomial of $\cos\theta$ with maximum exponent $\leq 2L$..

(ii) ϕ_1 has a definite parity which is the opposite of that of ϕ_0. Thus the interference term between ϕ_0 and ϕ_1 gives rise to odd powers of $\cos\theta$ only and we have the result that the terms in the angular correlation to the first order of $\frac{v}{c}$ is an odd polynomial of $\cos\theta$.

The author wishes to take this opportunity to thank Prof. E. Teller for invaluable discussions and advice.

ABSTRACT

Theorems concerning the general form of the angular distribution of products of nuclear reactions and disintegrations are derived. These theorems are based only on the invariance properties of the physical process under space-rotation and under inversion. The following examples are studied in detail: (i) Angular correlation between the electron and the neutrino in β-decay. (ii) Angular correlation between a β-ray and a γ-ray emitted in succession by a nucleus. (iii) Angular correlation between two γ-rays emitted in succession by a nucleus.

杨振宁在国立西南联合大学期间的学业

一、物理学系本科[1]（1938—1942）

1. 公共必修课

课程	授课教师 *	成绩 **
大学一年级（1938—1939 学年）		
国文读本（两学期，4 学分）	朱自清	78，（），平均 78
国文作文（两学期，2 学分）	朱自清、闻一多、罗常培、王力等，每人讲授一至两周	74，77，平均 76
英文一（两学期，6 学分；读本 3 周学时 + 作文 2 周学时）	叶公超	80，83，平均 83
中国通史（两学期，6 学分）	钱穆，雷海宗	98，95，平均 95
经济学概论（两学期，6 学分）	陈岱孙，王秉厚（?）	87，82，平均 85
大学二年级（1939—1940 学年）		
德文一（两学期，6 学分）	冯承植，陈铨，杨业治，雷夏（?）	96，（），平均 98

注：此外，还有体育、军训和党义等课。体育课每学期都有，见成绩单。

* 此处所列授课教师主要根据杨先生回忆。部分课程的授课教师杨先生不能完全确定的，本文引用参考文献所列授课教师名单，并以（?）表示存疑。如在 1939—1940 学年讲授德文一课程的四位授课教师，其姓名来自参考文献，不是杨先生回忆的，实际上给杨先生上课的教师可能是其中某位或某两位。后同。

** 课程为两学期的，所列分数分别为上、下学期成绩和记入学分积的全学年平均成绩，（）表示成绩单上该学期成绩空缺。后同。

2. 专业必修课

科目	授课教师	成绩
大学一年级（1938—1939 学年）		
微积分（两学期，8 学分）	姜立夫	95，97，平均 96
普通物理（两学期，8 学分；包括普通物理实验，基本上每周一次，不另计学分）	赵忠尧	95，95，平均 95

---

[1] 杨振宁高中阶段没有学习过物理，1938 年，他以同等学力考入西南联大化学系，是以成绩单上记载，"民国 27 年 11 月考入理学院化学系"。入学前杨振宁读了物理教科书，产生兴趣，一入学就转到物理学系。由于西南联大一年级实施通识教育，理学院课程设置均相同，物理学系唯一的要求是大一必修普通物理且成绩满足要求。所以他的成绩单上大学一年级的成绩记录为"化学系第一年级"。

续表

科目	授课教师	成绩
大学二年级（1939—1940 学年）		
高等微积分（两学期，8 学分）	曾远荣	98，100，平均 100
力学（两学期，6 学分）	周培源	90，95，平均 93
电学（两学期，6 学分）	吴有训	100，96，平均 98
高等物理实验一（电磁学实验，3 学分）	虞福春	80
普通化学（两学期，6 学分）	孙承谔	95，89，平均 92
普通化学实验（两学期，2 学分）		85，（ ），平均 91
大学三年级（1940—1941 学年）		
微分方程论（下学期，3 学分）	陈省身	96
热学（两学期，6 学分）	叶企孙	95，95，平均 95
光学（两学期，6 学分）	饶毓泰	90，90，平均 90
高等物理实验二（光学实验，1.5 学分）		85
无线电原理一（上学期，4 学分）	任之恭，孟昭英	94
无线电实验一（上学期，1.5 学分）		85
大学四年级（1941—1942 学年）		
近代物理（两学期，6 学分）	吴有训	82，93，平均 88
近代物理实验（下学期，1 学分）		78
无线电原理二（上学期，4 学分）	朱物华	90
无线电实验二（上学期，1.5 学分）		85
微子论（上学期，3 学分）	马仕俊	95

3. 选修课

科目	授课教师	成绩
大学三年级（1940—1941 学年）		
物性论（上学期，3 学分）	张文裕	91
德文二（两学期，6 学分）	杨业治	94，95，平均 95
英文二（两学期，6 学分）	卞之琳，柳无忌，潘家洵，练北胜，谢文通，徐锡良，凌达杨(?)	84，85，平均 85
微分几何（上学期，3 学分；研究生课程，或算学系本科生课）	陈省身	93

续表

科目	授课教师	成绩
大学四年级（1941—1942 学年）		
流体力学（上学期，3 学分；研究生课程）	周培源	98
统计力学（下学期，3 学分；研究生课程）	王竹溪	67
理论物理（两学期，6 学分；研究生课程）	马仕俊	98，95，平均 96.5

二、清华大学研究院物理学部（1942—1944）

科目	授课教师	成绩
量子力学（研一整年，6 学分）	王竹溪	98，96
X 射线及电子（研一整年，6 学分）	吴有训	88
光之电磁论（研一上学期，3 学分）	饶毓泰	75
相对论原理（研一下学期，3 学分）	周培源	80
原子核物理及放射性（研一整年，6 学分）	张文裕	90

（杨振宁在国立西南联合大学学习期间的课程、授课教师、学分和成绩由朱邦芬整理，经杨振宁先生审阅。参考文献：王学珍, 江长仁, 刘文渊. 国立西南联合大学史料三（教学、科研卷）[M]. 昆明：云南教育出版社, 1998）

杨振宁的国立西南联合大学及芝加哥大学成绩单

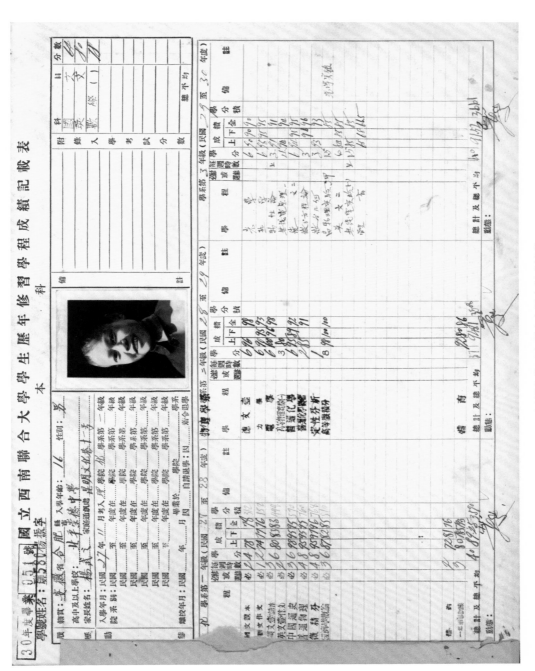

杨振宁的西南联合大学本科成绩单
（清华大学档案馆藏）

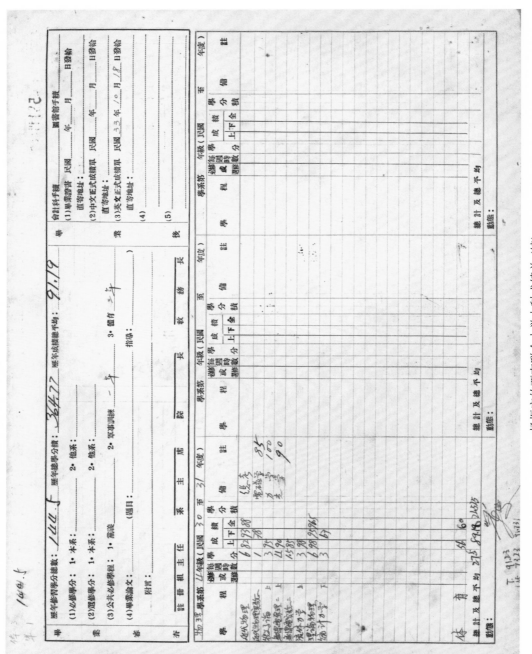

杨振宁的西南联合大学本科成绩单（续）
（清华大学档案馆馆藏）

國立清華大學 研究院

32年度畢業 2號

學號姓名：研248 楊振寧（男）
入學年度：民國31至32年度
入學考試總平均成績：%
籍貫：安徽省 合肥縣　入學年齡：十九
研究期滿年月：民國33年7月
以前畢業學校：國立西南聯合大學
授予學位年月：民國　年　月

科研究所研究生歷年成績學分配憶表
學部：

第一年（民國31至32年度）			第　年（民國　至　年度）			第　年（民國　至　年度）		
學程	成績 上學期 下學期 總平均	學分	學程	成績 上學期 下學期 總平均	學分	學程	成績 上學期 下學期 總平均	學分
量子力學	97 6						上學期 下學期	
原子物理	90 6							
統計力學	3							
相對論（旁聽）	80 3							
老實驗（旁聽）								
X Ray 晶子	182 6							

學年平均成績：88.13　24　學年平均成績：　　學年平均成績：

應州學分考試：成績及格　國考日期33年3月16日　應年修習學分總數：24　應年學分成績總平均：88.13%　按25%計 22.03%

畢業初試：成績：89%　舉行日期：三十三年七月十四日　按25%計 22.25%
應考學程：
考試委員：

論文考試：成績：88%　舉行日期：
　　　　　導師：
論文題目：Investigations in the statistical theory of superlattices
考試委員：
按50%計 44.00%

研究期滿考試成績：88.28%

備註：

楊振寧的清華大學研究院碩士研究生成績單
（清華大學檔案館館藏）

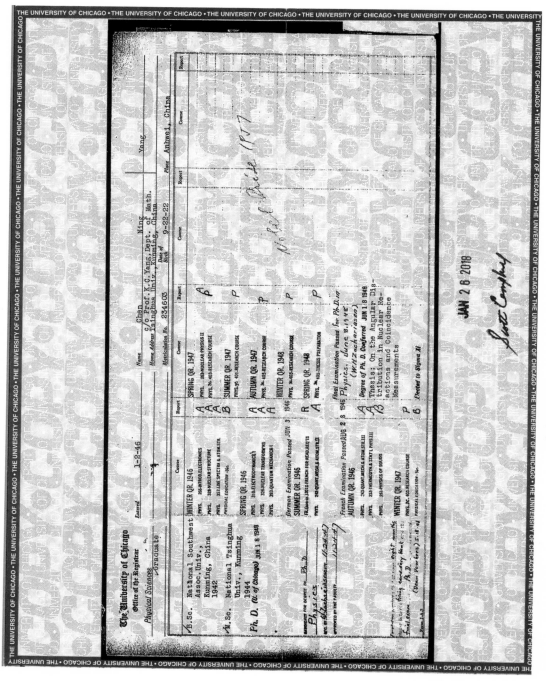

西南联大的人才培养和杨振宁先生的学术起步
——代后记

朱邦芬

在抗日战争的艰难环境中，由清华大学、北京大学、南开大学联合组成国立西南联合大学（以下简称西南联大）。西南联大的本科教学和研究生教学均由联大统一组织。杨振宁先生的学士学位由西南联大授予，学士论文导师是北京大学物理系吴大猷教授；硕士学位由清华大学授予，导师是清华大学物理系王竹溪教授。杨先生在许多场合多次提及西南联大的教育对他的影响。在《杨振宁文集》中，他写到："西南联大是中国最好的大学之一。我在那里受到了良好的大学本科教育，也是在那里受到了同样良好的研究生教育，直至 1944 年取得硕士学位。""我在西南联大 7 年，对我一生最重要的影响，是我对整个物理学的判断，已有我的 taste。"[1]（注：taste 中文译成品味或爱憎，杨先生对此两译名均不甚满意。以下，我有时用品味，有时直接英文。）2017 年 11 月 1 日纪念西南联大建校 80 周年时，杨先生又说："感谢西南联大，它是奠定我一切成就的基础。"

1938—1946 年期间，西南联大就读学生 8000 有余，毕业学生 3000 多人，在条件简陋、生活艰苦的情形下，西南联大培育了一大批杰出人才，包括 2 位诺贝尔奖获得者——杨振宁、李政道，4 位国家最高科技奖获得者——黄昆、刘东生、叶笃正、郑哲敏，6 位"两弹一星"功勋奖章获得者——郭永怀、陈芳允、屠守锷、朱光亚、邓稼先、王希季，数十位中国科学院和中国工程院院士，以及更多的中国各个领域的开创者和领军人物。

西南联大为什么育人成功？原因涉及很多方面，譬如：优中选优、出类拔萃的学生，国内超一流的师资，优良的学风，教授治校的体制，睿智的学校领导人，特别是"教育救国、科学救国"的强烈救亡意识和"刚毅坚卓"的精神与动力，第二次世界大战胜利及中华人民共和国成立以后百废待兴、急需人才的客观环境，等等。这些均已有很多研究。然而就杰出人才培育诸环节中最重要的一个环节——学生主动获取知识、科研训练和毕业论文——而言，特别是通过科研训练培育天才学生的 taste，从而为其成为世界级杰出人才的风格形成打下基础这方面，虽有一些很好的回忆文章，但深入细致的研究还比较少。本书收集了杨振宁先生从大学本科到博士的三篇学位论文——西南联大学士毕业论文、清华大学硕士学位论文和芝加哥大学博士学位论文，其中前两篇学位论文都直接与西南联大的培养有关。更为难得的是，杨先生专门为此撰写了评注，提供了他的三份成绩单（注：本科及硕士成绩单复印件是在 1992 年杨振宁先生 70 寿辰时，清华大学时任校长张孝文代表清华大学赠送给杨振宁先生的,芝加哥大学应杨先生要求于不久前提供了他博士期间的学习成绩单),并与我们多次深入交谈和通信,解答我们的疑问。杨振宁先生的西南联合大学本科成绩单和清华大学研究院硕士成绩单,原件保存在清华大学档案馆。承蒙清华大学档案馆大力协助,在庆贺杨振宁先生百岁寿辰之际,为本书慷慨提供馆藏成绩单的高清晰照片,使得复印件看不清楚的细节首次得以面世。这样,本书为大学教师、学生和管理人员,为教育研究工作者,特别是为关心中国杰出科学人才培养的各方面人士,提供了一位大师在大学成长的典型案例。

本文仅就西南联大物理系育人的理念和实践提供一个背景介绍,阐述西南联大的教育为杨振宁先生以后学术上的飞跃做了些什么样的准备;与此同时,对杨振宁先生三篇学位论文的背景和内容作一概要介绍。

西南联大物理系育人理念和实践

比较西南联大物理系的本科生和研究生的培养体制与 20 世纪 30 年代清华大学物理系的学生培养体制,可以发现两者十分相似。西南联大的三位常委,因蒋梦麟、张伯苓均在重庆任职,而清华梅贻琦长期在昆明,是实际上的校务主持人。联大物理系的师资队伍中,属于清华的超过一半(例如,1940 年联大物理系教授,属于清华教授有叶企孙、吴有训、周培源、赵忠尧、霍秉权、王竹溪、任之恭、孟昭英等 8 人,北大物理系教授有饶毓泰、朱物华、吴大猷、郑华炽等 4 人,南开物理系有教授张文裕 1 人),此外还有与物理有密切关系的清华金属研究所和无线电研究所的范绪筠、余瑞璜、叶楷等教授,再加上清华物理学的教学脱胎于美国,而西南联大物理系大多数教授有留美的经历,因此研究杨振宁先生在西南联大所受到的教育的理念和许多做法,可以追溯到早期的清华大学物理系的人才培养理念。

1926 年由叶企孙创建、抗战前由叶企孙和吴有训两位先生主持的清华大学物理系,其人才培养理念主要反映在以下这两段文字:

1927 年,在《清华物理学系发展之计划》一文中,叶企孙先生提出:"我们的课程方针及训练方针,是要学生想得透;是要学生对于工具方面预备得根底很好;是要学生逐渐的同我们一同想,一同做;是要学生个个有自动研究的能力;个个在物理学里边有一种专门的范围;在他们专业范围内,他应该比先生还懂得多,想得透。倘若不如此,科学如何能进步?"[2]

1931 年、1934 年和 1936 年,叶企孙在《物理学系概况》中三次写到:"在教课方面,本系只授学生以基本知识,使能于毕业后,或从事于研究,或从事于应用,或从事于中等教育,各得门径,以求上进。科目之分配,则理论与实验并重,重质而不重量。每班专修物理学者,其人数务求限制之,使不超过约十四人,其用意在不使青年徒废其光阴于彼所不能学者。"[3]

从叶先生上述两段有关培养学生的话,结合其实践,我们可以概括出老清华物理系和西南联大物理系教书育人理念的四个要点。

(一) 在课程设置和教学内容上"只授学生以基本知识"

20世纪20年代末至30年代中，中国的高等教育有比较大的发展，大学数量增多，许多大学新开设物理系，然而物理师资力量普遍较弱，物理教学还没有纳入正规，"科目过于繁多"，教材"华而不实""流于空泛"，且缺乏教学实验。老清华物理系教学上"只授学生以基本知识"，返璞归真，反而有利于学生个性化培养，使各类学生各得其所，不仅有助于学习一般的同学打下扎实基础，增强其自信心，还可以使优秀学生有更多的时间和精力去主动学习和主动研究，变接受教师传授知识为自己主动获取知识，而这对于天赋优异的学生尤为重要。

联大本科教学与老清华相同，学制为四年，实行学分制，要求修满132学分。课程分为三类：全校共同必修课、系所规定的必修课（包含限选课）和选修课。此外还有不计入总学分而要求必须及格的三门课：体育、党义和军训，如果不及格照样要重修。继承老清华重视体育的传统，体育课每学期都要上，如果某个学期体育不及格，将可能因补修而推迟毕业。

按照梅贻琦先生的理念，本科教育应以"通识为本，专识为末""大学期内，通专虽应兼顾，而重心所寄，应在通而不在专"。七七事变以前，清华师生曾就大学通识教育的地位和重要性发生过不同意见的公开争论，最后妥协的结果是，本科通识教育主要在大学一年级进行，文、理、法三个学院一年级课程设置都一样，而工学院大一课程设置略有不同。

联大实行的"通识教育"与老清华相同。从杨振宁本科成绩单可以看到，联大理学院一年级"通识教育"的5门公共必修课分别是：(1) 国文（两学期共6学分，其中读本4学分，作文2学分）；(2) 英文（两学期共6学分，其中读本3学时/周，作文2学时/周）；(3) 中国历史或西洋历史（杨振宁选了中国历史，两学期6学分）；(4) 微积分、逻辑或高等数学（杨振宁选了微积分，两学期8学分）；(5) 普通物理、普通化学、普通生物、普通地质学（物理系要求本系学生大一必修普通物理课，两学期8学分，每周一次2~3小时实验，不另计学分）。联大还要求学生在大二或高年级从经济学概论、政治学概论和社会学概论3门课中任选一门作为公共必修课，杨振宁选修了经济学概论（6学分）。这样，杨振宁在联大第一年上了40学分的课。

联大学生从大二起选修第二外国语等公共必修课（德文或法文，杨振宁大二修了两学期德文，6学分）。从大二开始，杨振宁修完物理系规定的必修和限选的专业课程有：力学（6学分），电学（6学分），热学（6学分），光学（6学分），近代物理（6学分），无线电原理（8学分），微子论（即气体运动理论，3学分），普通化学（6学分），高等物理实验一（电磁学实验，3学分），高等物理实验二（光学实验，1.5学分），无线电实验（3学分），近代物理实验（1学分），普通化学实验（2学分）。连同大一的普通物理（包括实验）和微积分，杨振宁在本科阶段所上的专业必修课共计有10门理论课和6门实验课，共计73.5学分。

杨振宁在联大选修的课程，除了4门研究生课外，还包括算学系的课程，高等微积分（8学分），微分方程论（3学分）；物理系选修课，物性论（3学分）。大三他还选修了德文二（两学期，6学分）和英文二（两学期，6学分），选修课共计41学分，其中专业选修课29学分。

杨振宁四年本科总学分是 144.5 学分，其中专业课 102.5 学分；大一和大三两学年都上了 40 学分的课，大二是 37 学分，大四由于撰写毕业论文，上课学分是 27.5。杨先生所修的总学分数超过联大规定的 132 学分，此外，他还完成了学士毕业论文。

确实，当年老清华和联大物理系所规定的必修课程，都属于物理学最基本的课程，数量比较少而且程度也比较浅，大致与美国大学物理系的本科教学相仿，完全没有今天国内大学多数物理系本科生必修的"四大力学"。虽然与今天清华大学物理系所规定要修的学分数相比要低一些，然而这只是最低标准，含金量并不低。对于像杨振宁这样学有余力的优秀学生，清华和联大物理系本科教学远远不限于此。

（1）虽然课堂上所讲授的内容比较简单，采用的教材比较简明，但许多物理和数学教师同时推荐 1~2 本有特色、难度较大的参考书，鼓励有兴趣的优秀学生进一步课余去自学。例如，大二杨振宁的"电学"是吴有训教的，所选用的教科书是 L. Page 和 N. I. Adams 编著的 *Principles of Electricity*，比较浅显，但吴有训同时还推荐 J. Jeans 的 *The Mathematical Theory of Electricity and Magnetism* 给优秀同学自学。

（2）虽然联大物理系只授本科学生以基本知识，但鼓励优秀学生在三年级或四年级选学研究生课程。杨振宁在本科阶段选修了 4 门研究生课程：大三上选修陈省身先生讲授的"微分几何"（3 学分），大四上选修周培源先生讲授的"流体力学"（3 学分），大四下选修王竹溪先生讲授的"统计力学"（3 学分），马仕俊先生讲授的"理论物理"（两学期共 6 学分）。王竹溪的课把杨振宁引入到他一生最钟爱的领域之一——统计力学，马仕俊的课把杨振宁带到场论的最新前沿。总的来说，"只授学生以基本知识"并没有妨碍西南联大优秀学生主动学习更多更深的知识，而且这些优秀学生通过自主学习而掌握高深内容，为他们未来的独立研究打下很好的基础。

（3）由于必修课程相对比较容易，很多学生有时间和精力旁听其他院系的课程，这非常有利于学生形成跨学科的知识结构，而这些并不反映在成绩单上。例如，杨振宁旁听过数学系许宝騄先生的"数理统计"，这门课要学两个学期，杨振宁旁听了一个学期，其中一半时间学习了矩阵论，为他牢固掌握量子力学打下十分坚实的数学基础；还有一半时间学习了测度论。回忆在西南联大求学时的老师，除了吴大猷和王竹溪两位导师，杨振宁认为许宝騄和马仕俊是"两位对我特别有重要影响的老师"，而这两位老师上的课都不是要求杨振宁本科阶段必修的。杨振宁还花了一个夏天时间自学了 E. T. Whittaker 和 G. N. Watson 的 *A Course of Modern Analysis*，做了大量习题，熟练掌握了各种特殊函数。这些数学课程，加上陈省身先生的"微分几何"和"微分方程论"课程，再加上杨振宁自小受到的数学熏陶，使得他的数学基础丝毫不亚于数学系学生。

(二) 理论与实验并重

中国传统教育重书本知识而轻实践，而奠定近代科学的基石——科学实验，直到一个世纪前才在中国学校作为一门课程诞生。物理学是实验科学，叶企孙、吴有训、饶毓泰等老前辈清楚认识到物理实验教学在培养科学人才中所起的不可替代的作用。老清华物理系本科实验有 4 个特点：学时安排比较多，实验课开设的种类也比较多，对实验课的要求很严格，实验课由名师担任。为了使学生的动手能力和实验技能得到切实的提高，学校为低年级学生开设木工及金工训练，"实验物理的学习要从使用螺丝刀开始"（吴有训语），"我们物理系是不给学生用好仪器做实验的"（叶企孙语）[2]。这样，同学从实验课程收获的，远超过在助教安排好的实验桌上做一些测量工作所得到的。老清华物理系还通过课外科研和毕业论文环节培养高年级学生的实验本领，此外还设有工场，招聘和培养高级实验技师，带领学生自制仪器，培养了学生的实际科研能力。例如，1935 年熊大缜在大学四年级研制出中国第一台红外照相机，在清华气象台拍摄了十分清晰的北京西山夜景，成为我国历史上第一张红外照片。

七七事变后，三校仓促内迁，虽然清华物理系在抗战前曾南运一批仪器，以后辗转运到昆明，但设备损坏丢失甚多，而有的大学实验仪器丧失殆尽，一时间实验课程难以开设。1938 年，利用中华教育文化基金会董事会补助的理工设备费 10 万元，联大物理系从上海和国外购得一批仪器设备，经越南海防和滇越铁路陆续运至昆明。利用这些设备，1939 年，西南联大物理系按萨本栋所著的《普通物理学实验》开设了为期一学年每周一次的普通物理实验，以后还陆续开设电磁学实验，光学实验，无线电实验和近代物理实验（6 个实验），基本恢复了战前清华物理系的实验课程。杨振宁在大学四年学习期间，每学期都有一门实验课程。除"力学"和"微子论"两门课外，其他各门物理课程均设实验课程，在抗战十分困难的环境中，十分难得。有一段时间，日本飞机经常空袭昆明，有的实验室每次做完实验，就把贵重仪器放进半埋在地下的 50 加仑的大汽油桶中，到下次做实验时再取出，以保证实验教学的正常进行。有的实验室（如无线电实验室），将仪器疏散到位于大普吉村的清华无线电研究所，需要时从乡间取来。

杨振宁的许多老师都是一流实验物理学家，如教"普通物理"的赵忠尧、教"电学""近代物理""X 射线及电子（研究生课程）"的吴有训、教"光学""光之电磁论（研究生课程）"的饶毓泰、教"热学"的叶企孙、教"物性论""原子核物理及放射性（研究生课程）"的张文裕、教"无线电原理"的任之恭和孟昭英。这些从事实验物理学研究的老师以及老清华物理系重视实验教学的传统的熏陶引起了杨振宁对实验物理的浓厚兴趣。虽然西南联大由于实验仪器和经费匮乏，很难开展真正的实验物理研究，但物理系学生可以花较多时间在实验室中重现物理学先贤们的发现，并认识到实验在物理研究中的重要地位。这些经历使得杨振宁始终重视实验研究在物理学发展中的极端重要性，并始终保持对实验物理的兴趣——即使他在实验方面不具有特殊才能。曾任美国布鲁克海文国家实验室主任的实验物理学家萨奥斯说过："杨振宁是一位极具数学头脑的人，然而**由于早年的学历**，他对实验细节非常有兴趣。他喜欢和实验学家们交谈，对于优美的实验极为欣赏。"这番话表明了西南联大物理系重视实验的教育对于杨振宁成长所起的积极作用。

(三) 重质不重量

对于老清华物理系强调的"重质不重量"，西南联大在昆明异常艰难困苦的环境中仍予以保持。

（1）课程由名师讲授。西南联大物理系的师资大多毕业于欧美一流大学研究院，均为当年国内一时之选。教师讲课不是单纯传授知识，而是在讲课时经常融入最新的科研进展和自己的体会，阐明科学大师的研究思路，也更注重启发学生思考。联大物理系荟萃三校教师精华，师资多且强，一门课程几位教授均可开设，每位教授能开多门课程，不少课程是由从事该学科研究的专家讲授的。给杨振宁授课的都是名师，科研与教学俱为上乘。除前面所述给杨振宁上物理课的实验物理学家外，还有许多杰出的理论物理学家给杨振宁上物理课，如周培源教"力学""流体力学（研究生课程）""相对论原理（研究生课程）"，吴大猷讲授"理论物理（研究生课程）"，王竹溪讲授"量子力学（研究生课程）""统计力学（研究生课程）"。教杨振宁数学课的教师也都是数学名师，如姜立夫的"微积分"，曾远荣的"高等微积分"，陈省身的"微分几何""微分方程论"，许宝騄的"数理统计"。此外，他的人文社会科学课程的教师也都是国内数一数二的大师，如讲授中文课的朱自清、闻一多、罗常培、王力，教英文的叶公超，教经济学的陈岱孙，等等。学生从良师"那里获得的东西中，不仅仅是知识或技能，最重要的是'思维风格'"。杨振宁先生曾说，"一个做学问的人，除了学习知识外，还要有 taste。一个人要有大的成就，就要有相当清楚的 taste"。影响 taste 的一个重要因素是"一个人在刚接触物理学的时候，他所接触的方向及其思考方法"。杨先生中小学没有上过物理课，而教他大学物理课的都是国内出类拔萃的名师，这对他形成自己独特的 taste 以至未来的研究风格而言，无疑极其幸运。

特别要指出的是，当时有一批青年教授——陈省身、王竹溪、马仕俊、许宝騄等，刚从欧洲获得博士学位回来，他们受到的训练、研究水准、对前沿的了解，总体上比西南联大资深教师更高，可以认为这批青年教授已经接近当时国际学科研究的前沿，而他们所讲授的课程也具有很高水准，把杨振宁直接带到了学科的最前沿，即使与同期国外一流大学物理系开设的研究生课程相比，也一点都不逊色。1945 年杨振宁到芝加哥大学物理系，他的物理和数学水平明显地高出他的同学一大截。正如杨振宁回忆："那几年我在昆明学到的物理已能达到当时世界水平。譬如说，我那时念的场论比我后来在芝加哥大学念的场论要高深，而当时美国最好的物理系就在芝加哥大学。""直到今天我还保留着王（竹溪）先生讲量子力学的笔记，它仍然是很有用的参考文献。笔记本的纸张很粗糙，没有漂白，很容易脆裂。每一次看到它，就会使我想起那艰难的岁月。"[1] 根据杨振宁的同窗好友黄昆回忆，虽然他名义上是硕士毕业，但实际水准，特别是量子力学的深入程度，已远远超过了他的英国同学，达到了博士后研究人员的水平。因而黄昆到英国"二战"后"最兴盛的理论学院"——布列斯托尔大学——读博士，觉得自己"基本知识增加很有限"。黄昆能够完全独立地开展研究工作，并在一年半时间内就完成博士论文，显然与他在西南联大所打下的基础密切相关。

（2）教与学的高标准严要求。虽然只授以学生基本知识，但联大教师教学认真，对学生要求严格，平时作业多，考试多，因此淘汰率比较高。正如杨先生回忆："总体来说我们的课程很有系统，准备充分，内容很有深度。"联大规定，公共必修课和物理系规定的必修课和限选课必须成绩合格；不及格者，不得补考，"必须重读"。杨振宁在西南联大是有名的学习成绩优异的学生，他大学四年的学分积高达 91.29 分，然而他成绩单上仍有个别课程成绩不高。例如，他在大四第二学期选修王竹溪讲授的研究生课程"统计力学"，成绩只有 67 分，这表明联大教师对学生的严格要求，绝不"放水"。

(3) 人才培养质量重于数量。联大物理系十分注意限定招收学生人数，把培养学生的质量放在首位。从 1938 年至 1946 年的 9 年中，联大物理系总共毕业本科生 130 人，一般每年毕业十余人。不仅限定本科生数量，研究生更是如此。在昆明期间，物理系一共招收研究生新生 12 人，其中清华研究院 9 人，北大研究院 3 人；即使招生最多的 1941 年和 1942 年，每年也只有 4 人，有的年份甚至为零。此外，清华还有 2 位复学的老研究生——钱伟长和谢毓章[2]。整个联大期间，物理学科在学研究生总共只有 14 人，而毕业研究生只有区区 7 位。这中间虽然有环境不安宁、部分研究生在学期间肄业、出国留学等因素，但联大限定招生数量和严把质量关是最主要的原因。

(四)"要学生个个有自动研究的能力"

西南联大物理系之所以培养了一大批优秀人才，其中一个关键点是"要学生个个有自动研究的能力"。尽管并非所有联大物理系本科生毕业以后都从事物理学研究，然而，旨在培养学生个个有自动研究能力的理念，使得学生在学习知识和探究知识过程中对新东西养成思考和分析的习惯，对探索未知领域产生浓烈的兴趣，进而使学生获得终身受益的探究能力和科学素养。这些对学生的未来成才十分重要。

吴大猷先生在《早期中国物理发展的回忆（续一）》中写到："我们该用怎样的标准来评估一个机构或是一些人对中国物理发展的贡献呢？主要是根据他们在若干年之内，是否有建立传统，包括人、设备与稳定的气氛等三方面；他们在几年内又能够吸引多少学生或是激励、唤起（inspire）多少个学生继续作物理研究工作。"（吴大猷，《物理》34 卷，4 期，236 页，2005 年）培育学生自动研究的能力，对于建立传统，对于吸引、激励、唤起学生继续做科学研究，无疑是最主要的一个渠道。

像杨振宁这样的天才学生，通过自动研究物理，形成了自己对于物理的 taste。杨振宁回忆："想起在中国的大学生活，对西南联大的良好学习风气的回忆总使我感动不已。联大的生活为我提供了学习和成长的机会。我在物理学里的 taste 主要是在该大学度过的 6 年时间里（1938—1944）培养起来的。"[1] 杨振宁"对物理学中某些方面的偏爱"，如对称性在物理学中所起的支配作用，多粒子系统的统计力学，就是他在昆明岁月里自主学习和自主研究中形成，并成为他一生最钟爱和最主要的两个研究领域。

为培育学生具有自动研究的能力，老清华和联大物理系有以下一些成功的做法。

(1) 给予学生自主的空间和充分的自由。清华物理系 1933 年毕业的傅承义院士回忆道："那时，我们并没有多少必修课，也没有做习题的压力，可以说是自由得很。大部分时间都是用来自学，凭着自己的兴趣钻研老师指定的参考书。"[2] 联大继承了这种自由传统。当人们问起"为什么西南联大当时的条件那么艰苦，却培养了那么多人才？"许多联大学生的回答往往是两个字——自由。

中国教育历来重视因材施教，但对于优秀学生，因材施教的传统做法是让他们"学得早一点，学得深一点，学得多一点"。但是，这样的因材施教存在一个基本缺点：对于优秀学生和学得一般的学生，同样的课程和大纲只有难易程度的差别，传授知识方式都是老师教学生学，而没有获取知识方式上的差别。杨振宁先生指出，"80 分以下的学生，在一个训导为主的教育体制下,成功的可能性比较大;90 分以上的学生,主要应该给他们启发性的鼓励,给自主的空间,让他们的主观能动性发挥得更好"。跟老清华物理系相比,联大战时特殊的环境,使得优秀学生有更充分的自由和更多的时间讨论和探索感兴趣的问题。杨振宁和黄昆、张守廉——驰名联大的"三剑客",一段时间内,从教室到寝室整天形影不离,课余大量时间"泡"在茶馆里切磋学问,畅谈学术人生。杨振宁说:"根据我读书和教书得到的经验,与同学讨论是深入学习的极好机会。多半同学都认为,从讨论得到的比老师那里学到的知识还要多,因为与同学辩论可以不断**追问**,深度不一样。"这样的场景在联大绝不是个例。同学间自由自在的讨论、辩论乃至追问,是联大出大批优秀研究人才的关键之一。我们现在一些重点大学把必修课压得太多,学时安排得太紧,给优秀学生自主学习和自主钻研的空间很少,学生没有时间去"胡思乱想",善于解题但不善于提出问题,这可能也是近些年来我国很少出现世界级的杰出科学人才的一个原因。

(2)努力创造科学研究的条件。在 20 世纪 30 年代,清华物理系拥有仪器约值 11 万银圆,图书室有物理方面的书籍千余册,国外成套的期刊及现刊十多种,再加上学校图书馆的书籍及杂志,以致从哈佛和麻省理工学院(MIT)留学归来的任之恭认为清华"物理系的图书馆要比哈佛大学的更加完善一些"(《中国科技史料》19 卷,1 期,43 页,1998 年)。南迁昆明后,图书仪器设备损失极大,资源极其匮乏。尽管如此,本着"知其不可为而为之"的精神,联大物理系师生仍想方设法做一些科学研究。在实验研究方面,主要有赵忠尧用由北平带出来的 50 毫克镭进行中子放射性元素实验;吴大猷用由北平带出的光谱仪棱镜等,放在木架制的临时性柱形(三棱镜)光谱仪,研究晶体拉曼(Raman)光谱;还有一些物理系教师和学生利用与清华金属研究所和无线电研究所的密切关系开展研究,其中有的研究在国际上相当有影响。如:余瑞璜 40 年代提出 X 射线衍射数据分析的新综合法和由相对强度决定绝对强度的方法,仅 1942 年一年内独立在英国《自然》(Nature)周刊发表论文 2 篇;范绪筠研究固体物理的一些基本问题,被国际上广泛引用,等等[2]。

(3)拥有一批热爱科学研究的教师。要培养学生具有自动研究能力,教师必须对研究有浓厚兴趣。七七事变以前,清华物理系教师们都有自己的科研课题,而且都是物理学当时的前沿热点问题,工作十分勤奋。例如,赵忠尧教授和助教傅承义在铅砖围着的实验台上用自制盖革计数器研究伽马射线,学生们都学会了制造和使用盖革计数器,而这在当时是研究放射性的必要工具。到昆明以后,鉴于实验研究需要较多的仪器设备和经费,西南联大物理系的研究主要以理论研究为主,如:周培源关于广义相对论和湍流的理论研究;王竹溪除了在统计力学领域的系统研究,还与汤佩松合作在生物物理领域最早运用热力学化学势来分析细胞内外水分的运动;马仕俊关于介子和量子场论的研究;吴大猷的英文专著《多原子分子结构及其振动光谱》(*Vibrational Spectra and Structure of Polyatomic Molecules*)在很长时间内是本领域的重要参考书[4]。这批热爱科研的教师的身教激励了学生热爱科学研究,成为他们的榜样。

(4)鼓励高年级学生开展科学研究。通过开展科学研究,可以使高年级大学生特别是研究生,在他们所从事的物理学特定专业范围内"比先生还懂得多,想得透"。在西南联大时期，物理系教师独自或指导学生在国内国外学术期刊共发表了 108 篇论文，此外还有多篇没有发表的论文，其中相当一部分是研究生和高年级本科生参与的。在十分艰难的战争时期，中国物理学会的全国年会仍然分区进行，1942 年中国物理学会第 10 届年会上，大学四年级的杨振宁参加了昆明地区的分会，和其他几位研究生报告了自己的研究成果。

杨振宁的三篇学位论文

大学本科生的主要任务是学习知识，学有余力的学生可以尝试做点研究，而大学学士学位论文是本科生在结束一个学习阶段时的综合性学习和研究训练，旨在使学生学会运用所学知识进行一些探究，为进一步的学习和研究打基础。研究生在学阶段则不仅要进一步掌握学科前沿知识，还要开始把重点转移到创造知识上。学习知识和创造知识两个环节，既有联系，又有实质性的差别。本科毕业论文一般是导师给一个研究题目，然后学生在导师指导下一步一步地学做研究；而博士研究生一般在导师所指定的研究领域中学做有创意的研究，特别是优秀研究生要学会自己寻找研究题目，最后达到具有独立研究的能力。博士毕业论文反映了博士生的研究能力，特别是研究工作的创造性和独立性。对于有潜质的优秀博士，其论文选题往往还具有相当的自主性。硕士学位论文介于两者之间：优秀的硕士学位论文完全可与博士学位论文媲美，而平庸的硕士学位论文甚至还不如优秀的学士学位论文。

通常，就培育学生研究能力、品味和风格而言，导师起很重要的作用。导师的研究领域、学术品味、思维风格和洞察力，往往给学生带来深远的影响，甚至影响学生未来的贡献。杨振宁很幸运也很特殊：在学术起步阶段遇到吴大猷和王竹溪两位良师，分别指引他进入到他一生发生浓厚兴趣的物理学两个前沿领域，而这两个领域都是刚开始不久的新兴领域。吴先生指引他进入对称原理研究领域，王先生把他带入统计力学领域，而杨振宁一生最重要的研究就是围绕对称原理和统计物理展开的。可以说，西南联大为杨振宁未来的学术腾飞奠定了极好的学术基础。

(一) 学士学位论文

1929—1937 年，清华大学物理系要求四年级学生撰写毕业论文一篇（相当于 3~4 学分），否则不能获得学士学位。七七事变后，不同于战前清华学士论文要求，限于研究条件，联大物理系规定本科毕业论文非必修，而由学生自选[4]。杨振宁选择了做学士毕业论文，论文题目是 *Group Theory and the Vibration of Polyatomic Molecules*，指导教师是吴大猷教授。杨先生提供的这篇学士论文是用打字机打在一种质地很差、很薄、容易破碎的纸上，有些字现在已看不太清楚了，公式和希腊字母都是杨振宁手写的。杨振宁的学士论文，除了学校保存一份，本人保留一份，根据吴先生的回忆，他保留了杨振宁的论文 30 余年，后来杨振宁告诉他，自己的一份已丢失，于是吴先生就把自己所保存的那份给了杨先生。

（1）论文题目选择

对于学位论文而言，论文题目的选择十分关键。学士学位论文的题目一般由导师布置。由于论文通常安排在大学四年级最后一个学期，需要在比较短的规定时间内完成，而此时学生往往还要上课；另外，选题还要有一定的新意，即使还不能发表在学术期刊上，但至少要在某几点或某一方面是以前没人做过的。因此，论文导师需要在了解学生能力的基础上，在自己所熟悉的研究领域给学生出合适的题目。当然，学生也可以自己选择毕业论文的题目。如果学生自己选题，要想得到导师的指导，题目应与导师的研究领域有较密切的联系。因而，不管学生自己选还是老师给题目，学生选择论文导师往往意味着自觉地或不自觉地选择了自己论文的研究领域，而其影响往往可能比想象的更为长远。

作为西南联大一名最优秀的大四学生，杨振宁的学士学位论文的题目是这样确定的。

根据吴大猷先生回忆，1941年，在他所教的古典力学、量子力学班中，"有杨振宁、黄昆、张守廉、黄授书、李荫远和其他十余人，遇见这样的'群英会'，是使教师最快乐的事，但教这样的一班人，是很不容易的事。除了我比他们多知先知一点外，他们的能力是比我高的。""在古典力学课将结束时，我出了十余个课题，任各人选一题做一篇。杨振宁选的是用群论方法于多原子的振动的问题。杨自力地读群论，读我给他的参考文章，写了一篇论文"。（注：在西南联大的相关教学资料中，1941—1942学年里，吴大猷只教过"量子力学及原子光谱"，没有吴先生回忆的"古典力学"，杨振宁成绩单上也没有这门课的记录。）西南联大期间，吴先生在极端艰苦的环境下撰写了一本书名为 *Vibrational Spectra and Structure of Polyatomic Molecules* 的英文学术专著。这本书出版后很快成为国际上多原子分子结构和振动光谱领域的权威著作。杨振宁的学士论文涉及吴先生最擅长的研究领域。另一方面，群论是杨振宁父亲杨武之所擅长的，30年代他曾在清华数学系讲授过群论这门课程，陈省身、华罗庚都曾听过。而杨振宁自己还在念高中时，"就从父亲那里接触到了群论的初阶，也常常被父亲书架上一本斯派赛（A. Speiser）的关于有限群的书中的美丽的插图所迷住"。

杨振宁先生回忆道："（吴大猷）给了我一本 *Reviews of Modern Physics*（《现代物理评论》），叫我去研究其中一篇文章，看看有什么心得。这篇文章讨论的是分子光谱学和群论的关系。我把这篇文章拿回家给父亲看，他虽不是念物理的，却很了解群论。他给了我狄克逊（Dickson）所写的一本小书，叫做 *Modern Algebraic Theories*（《近代代数理论》）。狄克逊是我父亲在芝加哥大学的老师，这本书写得非常合我的口味。因为它很精简，没有废话，在二十页之间就把群论中'表示理论'非常美妙地完全讲清楚了。"

（2）用群论于多原子分子振动谱

多原子分子的振动，是指分子中每个原子围绕其平衡位置所做的微小振动。原子在平衡位置受力为 0，而偏离平衡位置时受到其他原子施加的作用力是微振动的驱动力。一阶近似下，作用力正比于原子偏离平衡位置的相对位移，以及弹性力常数（即势函数对位移的二阶导数）。每个原子的位移有三个分量，即三个运动自由度，由 N 个原子组成的分子，一共有 $3N$ 个运动自由度。对于每个原子位移分量，可以写出一个由 $3N$ 个位移分量组成的线性齐次方程，这样的方程一共有 $3N$ 个。$3N$ 个线性齐次方程组存在非平庸解的条件是其系数矩阵行列式为零，由此可得到分子的本征振动频率和本征振动模式。在电脑和数值计算程序包广泛使用的今天，只要矩阵维度不太大，求解这样的久期方程不是一件困难的事。

然而，八九十年以前，计算条件的限制使得求解多原子分子振动频率和简正模式不是一件容易的事。另一方面，多原子分子的平衡位置一般都具有某种对称性（如空间旋转特定角度，镜面反射，中心反演，等等），利用点群对称性和群论的表示理论，可以大大降低久期方程的维度，由此比较容易求得多原子分子振动频率；还可以直截了当地观察到简正模式的对称性。吴先生给杨振宁的参考文献 *Group Theory and the Vibrations of Polyatomic Molecules*（注：杨振宁学士论文题目与之完全一样！），刊登在一本 1936 年出版的《现代物理评论》杂志上（Rev. Mod. Phys. 1936, 8(4): 317–346），作者是美国哥伦比亚大学化学系的 Jenny E. Rosenthal 和 G. M. Murphy。这篇参考文献共 30 页，但其中约 2/3 篇幅是从群的定义开始介绍群论的最基本知识，可见当年物理学界对群论是相当不熟悉的；甚至还有不少大物理学家公开表示讨厌群论，泡利（W. E. Pauli）甚至称之为"群祸"。在吴先生指导下，杨振宁选择用群论方法研究多原子分子的振动谱。杨武之提供的狄克逊写的《近代代数理论》这本书和吴先生提供的《现代物理评论》的这篇评述论文构成了杨振宁学士论文的学术背景。

如果只考虑分子内部原子间的相对运动，则应该去掉代表分子整体平移的 3 个自由度和整体空间转动的 3 个自由度，因此 N 原子分子的振动剩下 $3N-6$ 个内部振动的自由度，这需要用 $3N-6$ 个"独立约化坐标"（或称"内部对称坐标"，或"几何对称坐标"）来描述。通常，用初等方法，去掉整体平动和转动的自由度，得到与位能对称性相合的内部对称坐标，是相当不容易的。

杨先生学士学位论文显示出他具有令人惊讶的成熟的数学技巧、极佳的数学推演和证明能力，即使现在一些相当优秀的本科毕业生，如果没有上过群论课，也不易完全掌握杨先生的学士论文。事实上，除了杨振宁的天赋外，精通群论的父亲对他的指点，通过自学完全掌握群论的"表示理论"，通过旁听许宝騄先生的课而掌握矩阵理论，这些都是这篇论文中体现出来的"令人惊讶的成熟的数学技巧"的原因。这篇论文也显示出杨振宁重视数学、善于将数学和物理结合的理论物理学家的特色。

（3）杨振宁学士论文的后续影响

通过准备学士论文，杨振宁学通了群论，体会到群论对称性的美妙，自此他对物理中的"很妙"的各种不变性产生了极大的兴趣，导致他进入了对称与不变性（invariance），或叫做"对称原理"的研究领域。这对他以后的研究有决定性的影响。对称原理，即"对称决定相互作用"，是杨振宁一生最主要的研究领域，占他所有研究工作的 2/3，包括他与米尔斯（R. L. Mills）提出的非阿贝尔规范场理论和与李政道合作发现的弱相互作用中宇称不守恒定律。杨振宁学士论文对他产生的长远影响不仅影响了他选择的未来的研究领域，还影响到他的学术品味和学术风格的形成。

(二) 硕士学位论文

（1）清华研究院培育研究生概况

1942 年秋天，杨振宁考进清华大学研究院理科研究所物理学部攻读硕士研究生，导师是清华大学物理系的王竹溪教授。那时考入联大的本科生都算作联大学籍，而研究生学籍则依导师所在学校分别归属于清华大学、北京大学、南开大学三个学校的研究院，虽然所有课程学习和考试仍然在一起进行。研究生入学考试一般都是基础课程，如 1939 年清华大学研究院理科研究所物理学部考试科目共 5 门：国文、英文（作文及翻译）、微积分及微分方程、力学及电磁学、热力学及光学。那年报名三人，录取了一个人。

清华研究院成立初始，就对研究生严格要求。按当时规定，研究生学制是两年，要取得硕士学位，必须满足以下要求：（1）修满 24 学分的课程，而取得学分的最低成绩是 70 分，否则必须重修而没有补考之说；（2）通过毕业初试；（3）通过论文考试。毕业成绩的计算方法为：课程平均学分积占 25%，毕业初试成绩占 25%，论文考试成绩占 50%，满分为 100 分。只有历年学分平均成绩、毕业论文及毕业初试三者皆及格者，才给予研究院研究期满考试及格之证书，并授予硕士学位。由于对研究生要求严格，抗战环境艰苦，再加上一些研究生在学期间肄业出国留学，整个西南联大物理学研究生只有 7 人毕业。其中黄昆属于北大研究院，谢毓章、黄授书、杨振宁、张守廉、应崇福和杨约翰 6 人都属于清华研究院[4]。

除了本科阶段修过的 4 门研究生课外，杨振宁在研究生院第一年修了 5 门研究生课（共 24 学分），分别是王竹溪讲授的"量子力学"（一学年 6 学分），吴有训讲授的"X 射线及电子"（一学年 6 学分），饶毓泰讲授的"光之电磁论"（一学期 3 学分），周培源讲授的"相对论原理"（一学期 3 学分），以及张文裕讲授的"原子核物理及放射性"（一学期 3 学分）。根据规定，研究生学年平均成绩如果不满 65 分，则要被退学。西南联大期间，本科生必修课程所指定的教材大多是美国大学通用的，比较浅显；而教杨振宁研究生课的王竹溪、马仕俊等都是从英国留学回来，所用教材均为当时世界上程度最深的，给他打下了很好的基础。杨振宁研一平均学分积为 89.125。此外，研究生第二外语考试必须通过，杨先生二外选的是德文。

毕业初试有点像博士资格考试，但限定范围，且仅口试。考试范围由各学部分别对每一位研究生作具体规定，杨振宁毕业初试科目为量子力学、统计力学及电力学三门，而他的同学张守廉（导师周培源）则考量子力学、相对论及流体力学。学生毕业初试须在完成毕业论文以前举行，成绩以百分制计算，70 分以上为及格；60 分以上、不到 70 分，可以申请补考。毕业初始由多名教授组成的考试委员会主持。张守廉、杨振宁两人的毕业初试同时进行，由同一个考试委员会主持。赵忠尧先生为邀请毕业初试的考试委员专门向清华大学教务长潘光旦先生写了如下报告。

敬启者：

物理系研究生张守廉、杨振宁二君业已经过德文考试，成绩及格。兹定于五月十二日（星期五）下午二至五时在西仓坡四号举行第一次考试，考试委员拟聘请严慕光先生（玉龙堆十六号）、郑华炽先生、杨武之先生、叶企孙先生、吴正之先生、王竹溪先生及忠尧担任。张守廉君考试科目为量子力学、相对论及流体力学，杨振宁君为量子力学、统计力学及电力学。特此奉请贵处函邀各考试委员，并转办事处预备地点并届时略备茶点，为盼。专此即请

潘教务长大鉴

赵忠尧敬启

杨振宁毕业初试得89分。

毕业论文须先经研究导师认可，再由论文考试委员会主持答辩，决定是否通过答辩，并给出论文成绩。按规定，多名教授组成的考试委员会还必须有经教育部核准的校外人员参加。杨振宁论文考试委员包括吴有训（吴正之）、叶企孙、王竹溪、马仕俊、赵忠尧以及北平研究院物理研究所钱临照、黄子卿诸先生。图1是以梅贻琦校长名义发出的杨振宁毕业论文考试委员会邀请函，相当正式。杨振宁的硕士论文题目为：（一）The Dependence of Lattice Constants and Interaction Energy on the Degree of Order；（二）A Generalization of Quasi-Chemical Method in the Statistical Theory of Superlattice。

杨振宁论文考试得88分。他的研究生三项成绩按权重平均后的总分是88.28分，这是一个非常高的分数。根据清华研究院规定，研究生年度成绩在75分以上的就可以

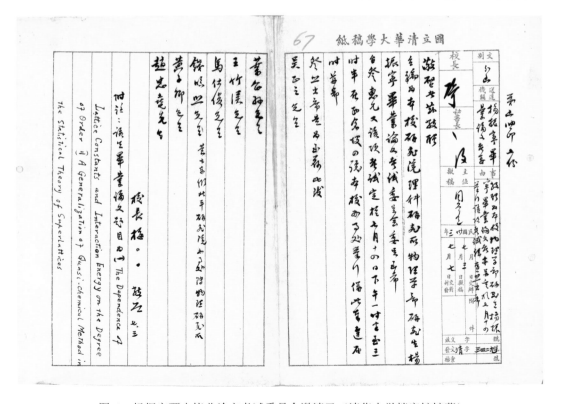

图1　杨振宁硕士毕业论文考试委员会邀请函（清华大学档案馆馆藏）

获得奖学金；如果毕业生的学分成绩、毕业初试成绩及论文成绩，均在 80 分以上，并获得所在学部推荐，可以由评议会特组审查委员会按照总成绩择优派遣留学，每年至多不得超过 10 名，每学部每年至多不超过 2 名。杨振宁的成绩完全够择优留学。当然，由于他又考取了清华庚款留美公费生，不需要评议会派遣留学。

（2）超晶格统计理论探究

杨振宁的硕士导师是王竹溪教授，硕士论文题目是《超晶格统计理论探究》。王竹溪先生于 1933 年清华物理系毕业后跟随周培源先生读研究生，1935 年在庚子赔款留英奖学金资助下去英国剑桥师从富勒（R. H. Fowler）教授，1938 年获得博士学位回到昆明任清华大学教授，专长是统计力学和热力学。杨振宁之所以选择王竹溪先生当他的研究生导师，一是因为王竹溪留英刚回来曾在联大围绕"相变"问题做过系列讲座，杨振宁都积极去听了，虽然似懂非懂但印象深刻，感觉"很妙"；二是王竹溪先生数学功底很深，很重视数学论证和物理规律的结合，这点很合杨振宁的 taste。在西南联大，王先生曾讲授过"普通物理""动力学""电动力学""量子力学""统计力学""微子论""热学"等多门课程，其中杨振宁选修过王先生教授的研究生课程"量子力学"和"统计力学"。王先生教学十分认真，学生中流传的一条重要经验是"谁要想学习理论物理学，一个最有效的办法是借阅王竹溪教授的笔记本看"。

金属合金由两种以上金属原子构成，按照原子排列是否有序，可把合金分为有序合金和无序合金。对于金属合金来说，最常见的无序是组分无序，即一种金属原子占据另一种金属原子的晶格位置。有序合金可以形成超结构，或称超晶格。一般而言，温度较高时，合金中不同金属原子的排列倾向于无序，而低温时比较有序，例如铜金合金有序时是四面体结构，在无序时晶格结构是面心立方，这样随温度升高，金属合金将经历由有序到无序的相变，相变时比热发生突变。20 世纪 30~40 年代，金属合金的有序、无序及其相变是统计物理学家和固体物理学家感兴趣的一个问题。富勒教授是该领域一位主要研究者，他的两位中国学生——王竹溪和张宗燧都开展过这方面的研究，而杨振宁关注这个问题的研究则是由王竹溪先生所引导。

杨振宁关于超晶格统计理论研究的硕士论文是由两篇研究论文组成：第一篇论文发表在《中国物理学报》（Chinese Journal of Physics）[1]上，标题为 Variation of Interaction Energy with Change of Lattice Constants and Change of Degree of Order（Chinese J. Phys. 1944, 5(2): 138–149）；第二篇论文发表在美国《化学物理杂志》（The Journal of Chemical Physics）上，题为 A Generalization of the Quasi-Chemical Method in the Statistical Theory of Superlattices（J. Chem. Phys. 1945, 13(2): 66–76），该文是杨振宁发表在国外的第一篇物理学研究论文。这篇文章编辑部收稿日期是 1944 年 11 月 17 日，寄出时间应该不会晚于 1944 年 10 月份。整个昆明期间杨振宁一共发表了 5 篇学术论文，其中发表在国外的论文 2 篇；还有一篇是数学论文 On the Uniqueness of Young's Differentials（Bull. Amer. Math. Soc. 1944, 50(6): 373–375），这是杨先生的第一篇国际论文。

---

[1]《中国物理学报》（Chinese Journal of Physics）创刊于 1933 年，发表的论文以英文为主，附中文摘要。1952 年改名为《物理学报》（Acta Physica Sinica），刊载中文论文。

铜原子和金原子组成的二元合金在热平衡时的位型由自由能极小决定，而自由能主要由原子的相互作用能和位型熵决定，其中相互作用能与有序度、晶格常数和原子排列有关。序一般可分为长程序和短程序；无序发生时，长程序首先被破缺，而某种程度的短程序仍保留。当时研究合金相互作用能和位型熵常用的方法有布拉格-威廉斯（Bragg-Williams）理论和贝特（Bethe）理论：前者不考虑短程序，而后者弥补了前者的缺点。杨振宁硕士期间第一项研究是用 Bethe 理论研究铜金合金的有序-无序问题，虽然以前已有人用 Bethe 理论研究了铜金合金的相互作用能，但在研究熵的时候却没有考虑短程序，杨振宁认识并解决了两者之间的不自洽性，从而计算得到的相变温度时的突变比热与实验符合得很好。

杨振宁第二项研究是将富勒和古根海姆（E. A. Guggenheim）提出的准化学方法做了进一步推广和发展。富勒和古根海姆将二元合金 AB 中各种可能配对的最近邻原子（包括错位）连同其相互作用能看成若干种"分子"，将合金看成分子的集合，从而由所谓的准化学方法，求解合金的一些物性。和 Bragg-Williams 方法、Bethe 等方法一样，准化学方法也属于平均场近似，其自由能求解都涉及复杂的积分和计算。杨振宁在硕士论文中不仅把准化学方法推广到更大的原子团簇，可以一阶一阶地做高阶近似，并明显比其他不同版本的平均场理论要好；更重要的是，杨振宁发现一个勒让德（Legendre）变换，可以直接计算自由能而避免了数学复杂性。这是一个很大的实质性的改进。杨振宁还运用这方法研究了更复杂的 Cu_3Au 面心立方结构。

1944 年杨振宁考取清华庚款留美奖学金，于 1945 年 8 月赴美之前在联大附中教数学，期间除了学习，还就二元合金的有序-无序转变问题完成了两篇论文：C. N. Yang. *The Critical Temperature and Discontinuity of Specific Heat of A Superlattice*（Chinese J. Phys. 1945，6(1): 59–66）；C. N. Yang and Y. Y. Li(李荫远). *General Theory of the Quasi-Chemical Method in the Statistical Theory of Superlattice*（Chinese J. Phys. 1947, 7(2): 59–71）。

（3）杨振宁硕士论文的后续影响

在王竹溪先生的引导下，通过硕士论文，杨振宁进入了统计力学研究领域。统计力学是杨振宁一生中另一个最主要的研究领域，大约占他所有研究工作的 1/3，他的 13 项最重要的物理学贡献中，4 项属于统计力学，包括相变理论、玻色子多体理论、杨-巴克斯特（Yang-Baxter）方程、一维 δ 函数排斥势中的玻色子在有限温度下的严格解[5]。2003 年杨振宁全时回到清华任教，随着冷原子实验物理的兴起，耄耋之年的杨先生又集中在统计物理领域两个物理结构直接简单的模型（稀薄玻色硬球系统和一维具有 δ 函数排斥作用的多粒子系统）开展了卓有成效的研究，取得了多项重要成果。

杨振宁硕士论文的第二项研究成果，被他收入在 1982 年出版的 *Selected Papers With Commentary(1945—1980)*，并列为文选的第一篇论文。在该篇文章的评注《忆我在中国的大学生活》中，杨振宁在回忆他在西南联大艰苦生活的同时，特别提及"我在物理学里的 *taste* 主要是在该大学度过的 6 年时间里（1938—1944）培养起来的"[1]。西南联大时期的学习对杨振宁产生了长远的影响，不仅影响他所选择的未来研究领域，还影响他同时对数学之美的欣赏和对物理之美的追求。

(三) 博士学位论文

（1）博士论文题目的选择

在清华留美公费资助下，1946 年初杨振宁注册成为芝加哥大学物理系的博士生。芝加哥大学物理系是当时美国最好的物理系，杨振宁于 1948 年 6 月就获得博士学位，并随后被留下当教员（instructor）。然而他的博士阶段并非一帆风顺。

杨振宁"去芝加哥的主要原因是想跟恩里科·费米 (Enrico Fermi，1901—1954) 写一篇实验方面的博士论文"。一方面，当时杨振宁觉得实验能力对于中国的发展更加需要；另一方面，他最欣赏三位物理学家（爱因斯坦、狄拉克、费米）的研究风格[6]，而只有费米兼做理论和实验，也只有费米有可能接受他当学生。可是，芝加哥大学核物理研究所 1946 年还没有破土动工，费米只能去阿尔贡（Argonne）实验室的反应堆做实验，而由于保密原因，杨振宁不能进阿尔贡。因而费米推荐他先跟特勒 (E. Teller) 做理论工作。特勒是美国的"氢弹之父"，他的新想法非常多，对于核物理学、凝聚态物理学、宇宙射线等问题都非常有兴趣。特勒建议他用托马斯-费米-狄拉克（Thomas-Fermi-Dirac）模型与维格纳-塞茨（Wigner-Seitz）近似方法计算 Be 与 BeO 的 K-电子湮没几率问题，杨振宁很快算出结果，并做报告获得好评。但是他对自己所用近似方法得到结果的可靠性没有把握，始终没有写成文章发表[7]。杨振宁渐渐发现，他所喜欢的研究方法与特勒的不一样，于是开始自己找理论研究题目。

1946 年秋天，费米又介绍杨振宁去做核实验物理学家艾里孙（S. K. Allison）教授的研究生，参与建造一台 40 万电子伏的考克饶夫特-瓦尔顿（Cockroft-Walton）加速器。可是杨振宁的动手能力较差，同学们很佩服他的理论知识，却一致笑他在实验室里的笨手笨脚，流传 "Where there is Bang, there is Yang!"

正如杨振宁在他的一篇介绍自己学习和研究经历的文章中所写，"博士生为找题目感到沮丧是极普遍的现象"[8]。1947 年，尽管杨振宁在《物理评论》(*Physical Review*) 已发表来美后的第一篇论文，《关于量子化的时空》(*On Quantized Space-Time*)，但是他在给好友黄昆的一封长信中曾用 "disillusioned"（幻灭）来描述自己的心情[9]。杨振宁学习一直极其优秀，自视甚高，又很努力，但是发现自己不擅长做实验，而做理论自己找的 4 个研究题目都一时没有取得成果。他当时发生兴趣的理论题目包括：（1）1944 年昂萨格（L. Onsager）关于伊辛（Ising）模型的文章；（2）1931 年贝特（H. Bethe）关于自旋波的文章；（3）1941 年泡利的关于场论的综合报告；（4）1943 年后关于核物理中各种反应的角分布的研究。受王竹溪先生的影响，前两个题目属于统计力学领域，受吴大猷先生的影响，后两个题目与对称性分析有关。当时，前三个题目芝加哥大学没人有兴趣，他独自一人在图书馆中研读，弄清来龙去脉，每一项都花了几个星期的努力，都以无果而告终。"只有第 4 项是特勒极感兴趣的研究。当时这方面的理论论文很多，可是都不够严谨，我花了几个星期用群论分析'物理规律旋转不变'（invariance of physical laws under space rotation）的意义，得出了几个漂亮的定理，写成一篇短文。特勒很喜欢这篇文稿。"

1948 年春天，特勒主动来找杨振宁说:"你不必坚持一定写出一篇实验论文。你已写了理论论文，那么就用一篇理论论文来做毕业论文吧。我可以做你的导师。"杨振宁

"想了两天，决定接受他的建议。作了这个决定以后，我如释重负"。以此为基础，构成杨振宁的博士论文。

（2）核反应和符合测量中的角分布

以往在计算核反应产物的角分布和涉及 β 射线和 γ 射线过程中角关联时，经过繁杂计算后经常遇到许多项互相抵消的情况。有人猜想这里面有与具体作用机理无关、普适的原理在起作用。杨振宁在博士论文中证明了这点，基于物理过程在空间旋转和中心反演不变，他推导了一般的定理，并详细分析了若干具体物理过程，包括 β 衰变中电子和中微子的角关联，核放射 β 射线和 γ 射线的角关联，和级连两束 γ 射线的角关联[7]。

杨振宁一开始的论文只有区区几页，特勒觉得作为博士论文，它太短了。杨振宁补充了半整数角动量（即考虑粒子的自旋）和考虑高速粒子的相对论效应，文章增加了几页，然而特勒还是嫌论文不够详尽，有些不高兴。最终杨振宁的博士论文扩充到十余页。文章简洁，不含一点渣滓，这正是杨振宁的风格之一，反映了作者思路明晰、逻辑性强，一气呵成的特点。

杨振宁在 1948 年 6 月获得博士学位，而《物理评论》收到他这篇研究论文也是在 1948 年 6 月 9 日，于同年 10 月 1 日正式发表。文章发表后，其中给出的定理受到核物理界的广泛注意。

（3）杨振宁博士期间研究工作的后续影响

杨振宁在两年半时间内获得了博士学位。鉴于他的数学和物理基础在去芝加哥大学以前就明显高于一般博士的水准，他在这两年半期间主要是全力以赴从事研究，他的收获也极其丰硕。

对称原理代表 20 世纪下半叶物理学发展的主流，也是 20 世纪理论物理三大主旋律之一。杨振宁的博士论文是他进入对称与不变性领域的第一篇文章，紧接着发表的关于 π^0 的自旋的工作是他在此领域中第二篇文章，文章中仔细分析了场论中不变性的群论表示。"这两篇文章使我一跃成为用群论与场论分析对称的专家，那时此领域那时才刚刚开始。"能在一个领域刚开始时进入该领域对于一个年轻科学研究工作者而言是极幸运的。

几 点 感 想

杨振宁从 1938 年秋考入西南联大，到 1948 年夏获得芝加哥大学博士学位，历时 10 年。这 10 年学校生活为他以后的学术腾飞奠定了扎实的基础，也给了我们许多启示，感受颇深。

(一) 学术大师主要地不是老师在课堂上教出来的，关键是要为有潜质的天才学生营造一个好的"环境"，以利于他们脱颖而出

抗战时期的西南联大，物质条件异常艰苦，然而杰出人才辈出，这里最重要的是它是一个好的广义学校"环境"。总结叶企孙先生创建的老清华物理系和西南联大物理系

的成功实践和理念,一个好的学校环境可以包括六点要素:优秀学生荟萃以及他们之间的相互作用;良好的学风;优秀的教师以及教师对学生尽心培育;学生有自主学习和研究的空间;国际视野;较好的软件和硬件条件。除最后一点,西南联大符合各个要素,对优秀学子而言确实是个好的学校环境。此外,抗战期间,亡国的危机时刻更加激发学生的使命感和责任感;而西南联大远在昆明,远离政治中心,学生相对而言更有自由。这些也为杨振宁成为大师提供了"沃土"。

(二) 一流大师在培育一流杰出人才方面起特别重要的作用,除"传授知识,培养能力,塑造价值观"外,一流大师对学生学术品味和学术风格的形成,特别是指引学生选择未来有发展前景的研究领域,有不可替代的作用

对杨振宁一生最有学术影响的导师有吴大猷、王竹溪、费米、特勒等4位教授。吴大猷引导杨振宁进入对称性原理领域,王竹溪引导杨振宁进入统计力学领域,特别是在这两个领域刚开始发展的时候引入,成为杨振宁一生主要研究和取得重要成果的领域。杨振宁从特勒那里不仅学到很多丰富的原创性想法,而且杨振宁的博士论文题目来自特勒物理直觉的启发:角分布和对称性及群论有关。更重要的是,特勒把杨振宁从执着于做一个实验物理学家,转变为做一位理论物理学家。

杨先生曾写到:"在每一个有创造性活动的领域里。一个人的 taste,加上他的能力、脾气和机遇,决定了它的风格,而这种风格反过来又决定他的贡献。"杨振宁曾把自己的研究风格归结为 (D+E+F)/3,其中 D 代表 Dirac,E 代表 Einstein,F 代表 Fermi。在西南联大 7 年,他没有机会见到这三位大物理学家,而是从阅读他们的文章而欣赏他们的研究风格,即具有把一个物理概念,一种理论结构,或一个物理现象的本质提炼出来的能力,并且都能够准确地把握住其精髓。虽然杨振宁从阅读文章中猜想并欣赏这三位大物理学家的风格,而后真正与大师接触,他觉得自己的感觉大致是正确的。我曾在一篇文章中概括杨先生的学术风格为:对数学之美的欣赏和对物理之美的追求并存;独立;简洁[10]。这的确与他这十年的研究经历有关,特别是费米对他的影响。

(三) 研究生阶段没有取得成果的研究是宝贵的财富

博士期间杨振宁有很长一段时间是自己在探索研究方向,遇到一些挫折,其实这是博士阶段最有价值的经历。判断博士论文能否通过和博士学位是否授予,最重要的一个判据是具备独立研究的能力。所谓独立研究能力,关键是自己找到有学术价值并可行的研究课题。这通常是不断探索、不断碰壁的过程,而这正是博士阶段最有价值的锻炼。

杨振宁在艾里孙实验室一年多时间,并不算成功。但是这段经历并不是白费时间,对他也是一笔宝贵的财富,使他"从中了解到实验工作者的价值观与理论工作者不同,这一点影响了我以后的许多工作,最显著的是 1956 年的宇称可能不守恒的文章与 1964 年的 CP 不守恒的唯像分析(phenomenological analysis)"。

博士期间，杨振宁自己独立选择了 4 个题目，都是极有学术价值的课题，虽然短时内都没有获得满意结果，遇到了一些挫折，但后来都先后开花结果了。

例如，对于 Ising 模型，1949 年 11 月，他在普林斯顿偶遇路丁格 (J. M. Luttinger)，听说考夫曼 (B. Kaufman) 已经简化了昂萨格的方法。由于杨振宁在芝加哥大学曾花了数星期研究昂萨格 1944 年的文章，做了必要的准备工作，"很容易就掌握了昂萨格-考夫曼方法的要点"，最后吸收了新方法，就开花结果了。这个"兴趣 → 准备工作 → 突破口"的过程，杨振宁认为是"多半研究工作必经的三部曲"。

又如，杨振宁在芝加哥读博时期对把外尔 (H. Weyl) 电磁学之规范不变性进一步推广到其他相互作用（泡利有名的综述报告中提及）十分有兴趣，1947 年他写了 3 页题为 *Gauge Invariance and Interaction* 的手稿（图 2 所示为其首页）[11]。类比用

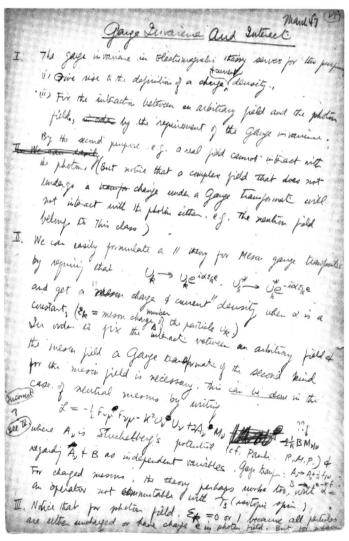

图 2　杨振宁 1947 年 *Gauge Invariance and Interaction* 手稿首页

矢量势描述、保持电荷守恒的电磁场方程，杨振宁试图导出用 B_μ 描述的、保持同位旋守恒的规范场方程，但后来总是导出一个越来越复杂的公式，迫使他不得不暂时搁置下来。他几次三番觉得这个想法很妙，但总是得不到满意的结果。正如杨先生在本书前言中提及，一直到 1953—1954 年他在访问布鲁克海文国家实验室期间，与同一办公室的米尔斯一起进行讨论，突然想到尝试在场强 $F_{\mu\nu}$ 上加一个二项式，变成 $F_{\mu\nu} = \frac{\partial B_\mu}{\partial x_\nu} - \frac{\partial B_\nu}{\partial x_\mu} + \mathrm{i}\varepsilon[B_\mu, B_\nu]$，从而一举奠定了杨-米尔斯（Yang-Mills）场的基础，成为标准模型的基石。

随着困难一一获得解决，对于选择研究课题，杨振宁自信心越来越强。他总结出"要找与现象有直接简单关系的题目，或与物理基本结构有直接简单关系的题目"，他还认为"把问题扩大往往会引导出好的新发展方向"，他多次以自身的经历告诫学生"最好在领域开始时进入一个新领域"。这既是他的经验也是他的研究风格之一。

参 考 文 献

[1] 杨振宁. 杨振宁文集 [M]. 上海：华东师范大学出版社，1998.

[2] 朱邦芬. 清华物理 80 年 [M]. 北京：清华大学出版社，2006.

[3] 叶企孙. 物理学系概况 [M]//清华大学史料选编. 北京：清华大学出版社，1991: 396。这段文字初载于 1931 年 9 月《清华消夏周刊迎新专号》，稍作修改后又载于 1934 年 6 月和 1936 年 6 月的《清华周刊 向导专号》。

[4] 西南联合大学北京校友会. 国立西南联合大学校史 [M]. 北京：北京大学出版社，2006.

[5] 施郁. 物理学之美：杨振宁的 13 项重要科学贡献 [J]. 物理，2014, 43(1): 57-62.

[6] 杨建邺. 杨振宁传 [M]. 上海：生活·读书·新知三联书店，2011.

[7] Max Dresden. 试论物理学中的风格和品味 [M]//丘成桐，刘兆玄，编. 甘幼玶，译. 杨振宁——20 世纪一位伟大的物理学家. 桂林：广西师范大学出版社，1993.

[8] 杨振宁. 我的学习与研究经历 [J]. 物理，2012, 41(1): 1-8.

[9] 朱邦芬. 读 1947 年 4 月黄昆给杨振宁的一封信有感 [J]. 物理，2009, 38(8): 575-580.

[10] 朱邦芬. 杨振宁先生的研究品味和风格及其对培育杰出人才的启示 [J]. 物理，2022, 51(1): 47-51.

[11] Chen Ning Yang. Selected Papers (1945—1980) With Commentary (2005 Edition)[M]. 新加坡：World Scientific Publishing Company, 2005.

编者简介

朱邦芬

  清华大学物理系教授，曾任清华大学物理系主任和理学院院长。2003 年当选为中国科学院院士，2012 年当选英国物理学会会士（Fellow）。曾任教育部物理类专业教学指导分委员会主任，中国物理学会副理事长，《中国物理快报》(Chinese Physics Letters) 主编，美国伊利诺伊大学、加州大学、香港科技大学等 8 所著名大学客座教授。现任《物理》主编，Solid State Communication、Chinese Physics B 等 5 种学术刊物编委，以及多个研究所和国家重点实验室学术委员。发表科学研究论文约 100 篇，编著书 8 本，曾获国家自然科学奖二等奖 2 项、中国科学院自然科学奖一、二等奖 3 项。朱邦芬是研究半导体量子结构物理的著名科学家，他与黄昆确立了半导体超晶格光学声子模式的理论，被国际学术界命名为"黄朱模型"，引起国际上普遍重视。他关注人才培养，是清华大学数理大类首席教授，教育部"基础学科拔尖学生培养试验计划"专家组成员，"清华学堂"物理班首席教授。曾获教育部第二届"杰出教学奖"，国家级教学成果奖一等奖，清华大学突出贡献奖等。

阮 东

  清华大学物理系教授，2007 年起担任清华大学物理系副系主任，2016 年起担任清华大学新雅书院副院长，2018 年起担任清华大学新雅书院党工组组长。现兼任教育部高等学校物理学类专业教学指导委员会副主任、中国高等教育学会理科教育专业委员会常务理事、中国物理学会物理教学委员会副主任、北京物理学会副理事长、全国群论教学研究会副理事长。曾获北京市高等教育教学成果一等奖 2 次，首届全国教材建设奖（高等教育类）二等奖，北京高校优秀本科教学管理奖，宝钢教育基金优秀教师奖，北京青年优秀科技论文二等奖，霍英东教育基金会高等院校青年教师三等奖等。